D1527064

# Physical Hazards of the Workplace

## Occupational Safety and Health Guide Series

### Series Editor

**Thomas D. Schneid**
*Eastern Kentucky University*
*Richmond, Kentucky*

### Published Titles

**Creative Safety Solutions**
*by Thomas D. Schneid*

**Occupational Health Guide to Violence in the Workplace**
*by Thomas D. Schneid*

**Motor Carrier Safety: A Guide to Regulatory Compliance**
*by E. Scott Dunlap*

**Disaster Management and Preparedness**
*by Thomas D. Schneid and Larry R. Collins*

**Managing Workers' Compensation: A Guide to Injury Reduction and Effective Claim Management**
*by Keith R. Wertz and James J. Bryant*

**Physical Hazards of the Workplace**
*by Larry R. Collins and Thomas D. Schneid*

# Physical Hazards of the Workplace

Larry R. Collins
Thomas D. Schneid

**Library of Congress Cataloging-in-Publication Data**

Catalog record is available from the Library of Congress



Direct all inquiries to CRC Press LLC, 2000 N.W. Corporate Blvd., Boca Raton, Florida 33431.



**Visit the CRC Press Web site at www.crcpress.com**


Lewis Publishers is an imprint of CRC Press LLC


International Standard Book Number 1-56670-339-5
Printed in the United States of America 1 2 3 4 5 6 7 8 9 0
Printed on acid-free paper

# Foreword

One stop shopping has become a mantra in our busy and recently computerized society. If you go to a "super store" you can find it all. If you go on the Internet, there are many sites that offer you access or portals to everything under the sun. But where does the new safety professional go to find a quick reference to the many critical regulatory issues facing him or her in their everyday professional lives? Dr. Larry Collins and Dr. Thomas Schneid have done an excellent job in this book of answering that question. *Physical Hazards in the Workplace* offers the student or novice in the field of safety, as well as the experienced professional, a quick reference guide to those issues that are dealt with daily or at least potentially faced on a regular basis.

Although we safety professionals spend a great deal of time dealing with or worrying about such hot topics as workplace violence and the proposed ergonomic standard, we cannot ignore those less glamorous issues that constantly create dangerous and/or deadly hazards and that, if not maintained daily (e.g., confined space, machine guarding, electrical safety), can seriously hamper if not destroy the work we have done to create preventative safety systems in our organizations.

The flow of this book is excellent for the student of safety. It begins with the regulatory issues to provide the student with an overview of what regulatory compliance is about and the nature of relationship of the government and industry. Specific reference is given to each area of physical hazard and the location where it can be found within the regulations.

The next chapter is a review of the currently hot topic of ergonomics. At the time of this printing, President Clinton had stepped in and used his influence and power to push through the final rule for an ergonomic standard. It is expected that industry will engage exhaustive litigation to fight this standard. One may ask, then, how does the final rule differ from last year's proposal?

Prior to the final rule, OSHA last proposed an ergonomics standard on November 23, 1999. Since that proposal was published, the agency received more than 8000 written public comments and listened to the testimony of more than 700 witnesses. Based largely on the public's participation in the process, the final standard contains significant changes. A fact sheet highlighting those changes is provided in an information kit, which is accessible on OSHA's website (www.osha.gov). One of the major concerns for industry is the expected annual costs of this standard. According to OSHA, they are estimated to be $4.5 billion, while the annual benefits it will generate are estimated to be $9.1 billion.

As outlined by the authors, the final rule uses a program approach for implementation of an ergonomic program that falls within the governmental standards. An acceptable program includes the basic elements of management leadership, employee involvement, job hazard analysis and control, training, musculoskeletal disorder (MSD) management, and program evaluation. These elements have previously been used in the red meat industry to successfully reduce MSDs. OSHA has used the basics of this ergonomics program as the foundation for the final rule.

As to coverage, the final rule will cover all general industry employers, including the U.S. Postal Service. OSHA estimates that 102 million workers at 6.1 million worksites will be protected. However, it should be noted that this new standard will not apply to employment that is covered by OSHA's construction standards (Part 1926), maritime standards (Parts 1915, 1917, 1918), and agriculture standards (Part 1928). It also does not apply to railroad operations. It will apply to public sector employers in OSHA's state-plan states if those states adopt the federal standard.

An interesting aspect of the new standard is that the "grandfather clause" of the proposed standard is still part of the final rule. Thus, for those employers who have already implemented ergonomics programs that contain the program elements of the final rule can continue their programs instead of complying with the requirements of this standard if their programs meet certain limited conditions and they include in their programs the MSD management that the standard requires.

In conclusion, the real benefit of this book is its easy-to-read and common-sense approach to these important issues. It combines the most common and important topics of our current safety concerns and provides a quick reference that can be used by anyone in the field of safety.

**Michael S. Schumann, Associate Professor**
Loss Prevention and Safety
Eastern Kentucky University

# *Preface*

Every year in America, many thousands of workers incur on-the-job injuries that cost them pain, suffering, and often economic hardship. Their employers face increasing costs to pay for workers' compensation, as well as the loss of production and quality, increased overtime by other workers, and a number of other costs because of the loss of injured workers. In virtually all circumstances, the injury or illness caused to a worker could have been prevented in the first place, with both the employer and the employee benefiting. Occupational injuries and illnesses can be prevented with the development and implementation of a proactive, well-managed, cost-efficient, and compliant safety and health program. This proactive effort can produce a "win-win" situation for all involved. In short, by preventing accidents, there are no losses. Spending money on prevention efforts pays dividends through reduced costs, increased production, and a happier work force.

In 1970, the Occupational Safety and Health Administration (OSHA) was established to safeguard the American workplace from known safety and health hazards. Several states have followed suit by establishing their own "state plan" programs. Over the years, OSHA has established a multitude of standards, or regulations, governing specific workplace hazard situations and methods to avoid workplace injuries or illness. Within this multitude of standards, some standards are very industry or situation specific, while others are very broad in their scope. Several standards address hazard situations that can create a greater risk of workplace injury or death, while others are aimed at injury or illness that can result from long exposure to a hazard. Several OSHA standards are consistently found to be violated year after year in the workplace for a myriad of reasons, including lack of knowledge about the standard or methods for achieving and maintaining compliance. These violations have cost employers millions of dollars in penalties and have created increased risk of injury to their workers.

In this text, we provide a basic education regarding several of the major regulatory issues, we address the physical hazards that have the greatest potential to cause on-the-job injuries, illnesses, or even death and the standards that apply to them, and we discuss many of the standards that have been identified by OSHA as being consistently violated in the workplace. Additionally, we examine several of the new and emerging issues impacting the American workplace and provide information and methods an employer can use to achieve and maintain compliance in a cost-efficient and effective manner. It is the authors' hope that this important information serves to assist companies, organizations, and workers in safeguarding their workplaces, removing physical hazards *before* accidents happen, and reaping the financial and personal rewards of their efforts.

# *About the Authors*

**Larry Collins**, Ed.D., joined the Loss Prevention and Safety (LPS) faculty of Eastern Kentucky as an Associate Professor in 1990. After serving as Program Coordinator of the Fire and Safety Engineering Technology Program, Dr. Collins assumed the role of Department Chair for the LPS department in 1998. Dr. Collins' background includes serving as a design draftsman for a tank semi-trailer manufacturer. He has 24 years' experience as a firefighter, having served in a large metropolitan fire department in northern Virginia and a combination career/volunteer fire department in Uniontown, PA. He has also been a fire instructor with the Pennsylvania Fire Academy. As BioMarine Industry's first fire service safety specialist, Dr. Collins traveled North America conducting training sessions on closed-circuit breathing apparatus and confined-space monitoring instruments.

Dr. Collins holds an A.S. degree in Fire Science from Allegheny Community College in Pennsylvania, a B.S. in Industrial Arts Education from California University of Pennsylvania, and a M.S. in Technology Education, also from California University of Pennsylvania. In July of 1993, Dr. Collins defended his doctoral dissertation, "Factors Which Influence the Implementation of Residential Fire Sprinklers," completing the requirements for his Ph.D. Dr. Collins is also nationally certified as a Vocational Carpentry Instructor. Dr. Collins is a strong advocate for residential fire sprinklers and believes this technology is the only real hope for changing the horrible fire death and injury statistics in the United States.

**Thomas D. Schneid,** Ph.D., J.D., is a tenured professor in the Department of Loss Prevention and Safety at Eastern Kentucky University and serves as the coordinator of the Fire and Safety Engineering program. He is also a founding member of the law firm of Schumann & Schneid, PLLC, located in Richmond, KY. Dr. Schneid has earned a B.S. in education, M.S. and CAS in safety, M.S. in international business, and Ph.D. in environmental engineering, as well as his Juris Doctor (J.D. in law) from West Virginia University and LL.M. (graduate law) from the University of San Diego. He has also earned an M.S. in international business and a Ph.D. in environmental engineering. He is a member of the bar for the U.S. Supreme Court, 6th Circuit Court of Appeals, and a number of federal districts, as well as the Kentucky and West Virginia Bars.

Dr. Schneid has authored and co-authored 15 texts on various safety, fire, EMS, and legal topics, as well as over 100 articles. He was named one of the "Rising Stars in Safety" by *Occupational Hazards* magazine in 1997 and was recently awarded the Program of Distinction Fellow by the Commonwealth of Kentucky and Eastern Kentucky University.

# Contents

**Chapter 1 Regulatory issues** ........ 3
*De minimis* violations ........ 8
Other or nonserious violations ........ 9
Serious violations ........ 10
Willful violations ........ 11
Repeat and failure to abate violations ........ 12
Failure to post violation notices ........ 13
Criminal liability and penalties ........ 13
OSHA inspection checklist ........ 18
Endnotes ........ 20

**Chapter 2 Ergonomics in the workplace** ........ 25
Implementing the program ........ 26
Management commitment and employee involvement ........ 26
Ergonomic team ........ 26
Worksite analysis ........ 26
Hazard prevention and control ........ 27
Engineering controls ........ 27
Workstation design ........ 27
Design of work methods ........ 28
Tool and handle design ........ 28
Work practice controls ........ 28
Administrative controls ........ 29
Personal protective equipment ........ 29
Medical management ........ 29
Periodic workplace walkthrough ........ 30
Symptoms survey ........ 30
Symptoms survey checklist ........ 30
Training and education ........ 30
General training ........ 31
Job-specific training ........ 31
Training for supervisors ........ 31
Training for managers ........ 31
Training for maintenance ........ 31
Auditing ........ 31
Disciplinary action ........ 32

Ergonomic checklist....................32
Proposed standard: ergonomics....................33
General industry....................33
Part 1910 [amended]....................33

**Chapter 3 Respiratory hazards**....................59
Respiratory hazard exposure prevention methods....................60
Classification of respiratory hazards: gases and fumes....................61
Different types of respirators available....................63
Open-circuit vs. closed-circuit SCBA....................64
Some thoughts on training respirator users....................65
Elements of a written respirator program....................67
1910.134(c)....................67
1910.134(l)....................68
1910.134(m)....................69

**Chapter 4 Fire and explosion hazards**....................71
Overall fire picture....................71
National fire problem....................72
Where fires occur....................72
Causes of fires and fire deaths....................73
Who is most at risk....................73
What saves lives....................73
Nonresidential properties....................74
Dealing with the fire problem....................74
Step one: assess your potential for fires and explosions....................74
Step two: maximize fire prevention efforts....................76
Step three: utilize early detection methods....................76
Step four: utilize automatic suppression systems....................76
Step five: slow the growth of potential fires....................77
Step six: prepare a fire pre-plan and train employees in it to minimize risk and maximize life safety and property protection....................77

**Chapter 5 Confined space hazards**....................79
Confined spaces and permit-required confined spaces defined....................80
Classification of hazards....................80
Respiratory hazards....................80
Corrosive hazards....................81
Energy hazards....................81
Engulfment (contents) hazards....................82
Fire and explosive atmospheric hazards....................82
Mechanical equipment hazards....................82
Fall hazards....................82
Environmental condition hazards....................83
Communication problems....................83

Atmospheric testing and monitoring ........ 84
Entry into a permit-required confined space ........ 85
  Permit-required confined space entry supervisor ........ 85
  Permit-required confined space entrants ........ 85
  Permit-required confined space attendant ........ 85
Confined space entry rescue ........ 86
Permit-required confined space rescue priorities ........ 86
  Priority one ........ 86
  Priority two ........ 86
  Priority three ........ 86
Permit-required confined space training ........ 87
Elements of a written confined space program ........ 88
Elements of a PRCS entry permit ........ 89

**Chapter 6 Electrical safety** ........ 91
Electrical safety ........ 91
General rules for operating around electricity ........ 92
  Employers ........ 92
  Employees ........ 92
Planning ........ 93
  General ........ 93
  Work plan ........ 93
  Work crews ........ 93
Safety equipment ........ 93
  Additional apparel considerations ........ 93
  Personal protective equipment ........ 94
    Protective helmets ........ 94
    Face and eye protection ........ 94
  Rubber goods ........ 94
    Testing and inspection ........ 94
    Storage ........ 95
    Use ........ 95
  Fiberglass and wood equipment ........ 95
Enforcement of the program ........ 95
Electrical safety checklist ........ 95
Control of hazardous energy (lockout and tagout) ........ 96
  Preparing for a lockout or tagout system ........ 97
  Sequence of lockout or tagout system procedure ........ 98
  Restoring machines and equipment to
    normal production operations ........ 98
  Training ........ 99

**Chapter 7 Machine guarding** ........ 101
Forming process actions ........ 103
  Cutting ........ 103
  Punching ........ 103

Shearing .......... 104
Bending .......... 104
What happens when the safety devices fail? .......... 104
Machine/equipment maintenance .......... 105
Personal protective equipment .......... 106

**Chapter 8 Fall hazards** .......... 107
Fall-protection categories .......... 111
Fall-arresting equipment .......... 111
Positioning .......... 111
Suspension systems and equipment .......... 111
Retrieval plan .......... 112
Fall-protection systems .......... 112
Duty to have fall protection .......... 112
Training requirements as set forth in 29 C.F.R. 1926.503 .......... 113
Certification of training .......... 114
Retraining .......... 114

**Chapter 9 Hearing protection** .......... 115
Hearing conservation .......... 117
29 C.F.R. 1910.95(a) .......... 117
29 C.F.R. 1910.95(b) .......... 118
29 C.F.R. 1910.95(c): hearing conservation program .......... 118
29 C.F.R. 1910.95(d): monitoring .......... 119
29 C.F.R. 1910.95(e): employee notification .......... 119
29 C.F.R. 1910.95(f): observation of monitoring .......... 119
29 C.F.R. 1910.95(g): audiometric testing program .......... 120
29 C.F.R. 1910.95(h): audiometric test requirements .......... 122
29 C.F.R. 1910.95(i): hearing protectors .......... 123
29 C.F.R. 1910.95(j): hearing protector attenuation .......... 124
29 C.F.R. 1910.95(k): training program .......... 124
29 C.F.R. 1910.95(l): access to information and training materials .......... 124
29 C.F.R. 1910.95(m): recordkeeping .......... 125
29 C.F.R. 1910.95(n): appendices .......... 126
29 C.F.R. 1910.95(o): exemptions .......... 126
29 C.F.R. 1910.95(p): startup date .......... 126

**Chapter 10 Bloodborne pathogens standard** .......... 127
OSHA bloodborne disease standard program development .......... 127
Example of a bloodborne pathogens program .......... 130
Introduction .......... 130
Instructions .......... 130
Bloodborne Pathogens Program .......... 131
Definitions (OSHA defined) .......... 131
Exposure Control Plan: overview .......... 132
General information .......... 133

Bloodborne Pathogens Manual ........ 134
Schedule of implementation ........ 134
Exposure determination ........ 134
Exposure determination lists ........ 135
Engineering and Work Practice Controls ........ 137
  Universal Precautions ........ 137
  Engineering and Work Practice Controls ........ 137
  Engineering Controls checklist ........ 138
  Inspection and maintenance schedule form ........ 138
Work Practice Control checklist ........ 140
Personal protective equipment ........ 141
  Availability ........ 141
  Gloves ........ 141
  Masks; eye and face protection ........ 141
  Protective clothing ........ 141
  Respiratory equipment ........ 142
  Cleaning, laundering, disposal ........ 142
  Replacement personal protective equipment ........ 142
  Attachment A to personal protective equipment ........ 142
Housekeeping ........ 145
  General ........ 145
  Equipment, environmental, and working surfaces ........ 146
  Glassware/reusable sharps ........ 146
  Regulated waste ........ 146
  Disposal ........ 147
  Attachment A to housekeeping ........ 147
Hepatitis B vaccine and vaccination series and post-exposure evaluation ........ 150
  Hepatitis B vaccination ........ 150
  Post-exposure evaluation and follow-up ........ 151
  Information provided to the healthcare professional ........ 151
  Healthcare professional's written opinion ........ 152
  Post-exposure response guidelines ........ 152
  Employees eligible for, and offered, hepatitis B vaccine and vaccination series ........ 154
  Hepatitis B Vaccine Declination Form ECP VIII-2 (mandatory) ........ 154
  Exposure incident checklist ........ 154
  Employee exposure incident record ........ 155
Hazard communications ........ 156
  Biohazard sign ........ 156
  Hazard communications checklist ........ 157
Training and education ........ 157
  Ems, Inc., training program requirements ........ 158
  Bloodborne pathogens training program ........ 158
Recordkeeping ........ 170

Medical records....170
Training records....170
Availability....171
Transfer of records....171
Recordkeeping checklist....171
Bloodborne pathogens program....172

**Chapter 11 Emergency and disaster preparedness**....175
Risk assessment....176
Planning design....177
Media relations....177
Remember the shareholder....179

**Chapter 12 Chemical hazards**....183
Employee Right-To-Know standard....183
History of chemicals in the workplace....183
Overview of the Hazard Communication standard (29 C.F.R. 1910.1200)....185
Standards cited for all SICs....186
Community right-to-know: a history of mistrust and disasters....187
Emergency Planning and Community Right-to-Know Act (42 U.S.C. 11001 et seq., 1986)....187
Emergency Planning and Community Right-to-Know Act mandates....188
Subchapter I — emergency planning and notification....188
Subchapter II — reporting requirements....189
Subchapter III — general provisions....190

**Chapter 13 Workplace violence**....191
Individual warning signs....191
Workplace violence prevention program....193

**Appendixes**

**Appendix A OSHA inspection checklist**....201

**Appendix B Employee workplace rights**....203
Introduction....203
OSHA standards and workplace hazards....204
Right to know....205
Access to exposure and medical records....205
Cooperative efforts to reduce hazards....205
OSHA state consultation service....206
OSHA inspections....206
Employee representative....206
Helping the compliance officer....206

Observing monitoring 206
Reviewing OSHA Form 200 207
After an inspection 207
Challenging abatement period 207
Variances 207
Confidentiality 208
Review if no inspection is made 208
Discrimination for using rights 208
Employee responsibilities 210
Contacting NIOSH 211
Safety and health management guidelines 211

**Appendix C Respiratory equipment** 213
Appendix A to §1910.134: fit testing procedures (mandatory) 213
Part I. OSHA-accepted fit test protocols 213
Part II. New fit test protocols 235
Appendix B-1 to §1910.134: user seal check procedures (mandatory) 235
Appendix B-2 to §1910.134: respirator cleaning procedures (mandatory) 236
Appendix C to §1910.134: OSHA respirator medical evaluation questionnaire (mandatory) 237
Part A. Section 1 (mandatory) 238
Part A. Section 2 (mandatory) 238
Part B 241

**Appendix D Confined spaces** 247
29 C.F.R. 1910.146 Appendix A: PRCS decision flow chart 247
29 C.F.R. 1910.146 Appendix B: procedures for atmospheric testing 249
Evaluation testing 249
Verification testing 249
Duration of testing 249
Testing stratified atmospheres 249
Order of testing 250
29 C.F.R. 1910.146 Appendix D: sample permits 250
Appendix D-1: confined space entry permit 250
Appendix D–2: entry permit 252

**Appendix E Fall hazards** 255
Inspection and maintenance 255
Harness inspection 255
Belts and rings 255
Tongue buckle 256
Friction buckle 256
Lanyard inspection 256
Hardware 256
Snaps 256

Thimbles ........ 256
Lanyards ........ 256
Steel lanyards ........ 256
Web lanyards ........ 256
Rope lanyard ........ 257
Visual indication of damage to webbing and rope lanyards ........ 257
Heat ........ 257
Chemical ........ 257
Ultraviolet rays ........ 257
Molten metal or flame ........ 257
Paint and solvents ........ 257
Shock-absorbing packs ........ 257
Cleaning of equipment ........ 257
Nylon and polyester ........ 258
Drying ........ 258

**Appendix F 29 C.F.R. 1910.1000: air contaminants** ........ 259
29 C.F.R. 1910.1000(a): Table Z-1 ........ 259
29 C.F.R. 1910.1000(b): Table Z-2 ........ 259
29 C.F.R. 1910.1000(c): Table Z-3 ........ 260
29 C.F.R. 1910.1000(d): Computation formulae ........ 260
29 C.F.R. 1910.1000(e) ........ 262
29 C.F.R. 1910.1000(f): Effective dates ........ 262
Table Z-1: Limits for air contaminants ........ 262
Table Z-2: Toxic and hazardous substances ........ 285

**Appendix G Workplace violence guidelines for night retail establishments** ........ 287
Introduction ........ 287
OSHA's commitment ........ 288
Extent of problem ........ 288
Fatalities ........ 288
Who are the victims? ........ 288
Nonfatal assaults ........ 288
Risk factors/summary of research ........ 289
Preventive strategies ........ 290
Overview of guidelines ........ 291
Violence prevention program elements ........ 291
Management commitment and employee involvement ........ 291
Written program ........ 292
Worksite analysis ........ 293
Records analysis and tracking ........ 293
Monitoring trends and analyzing incidents ........ 294
Screening surveys ........ 294
Workplace security analysis ........ 295
Hazard prevention and control ........ 296

Engineering controls and workplace adaptation ........ 296
Administrative and work practice controls ........ 297
Proper work practices ........ 297
Monitoring and feedback ........ 298
Adjustments and modification ........ 298
Enforcement ........ 298
Post-incident response ........ 299
Training and education ........ 299
General training ........ 299
Training for supervisors, managers, and security personnel ........ 300
Recordkeeping and evaluation ........ 300
Recordkeeping ........ 301
Evaluation ........ 301
Sources of assistance ........ 302
Conclusions ........ 302

**Appendix H Machine guarding checklist** ........ 305
Requirements for all safeguards ........ 305
Mechanical hazards ........ 306
Point of operation ........ 306
Power transmission apparatus ........ 306
Other moving parts ........ 306
Nonmechanical hazards ........ 306
Electrical hazards ........ 307
Training ........ 307
Protective equipment and proper clothing ........ 307
Machinery maintenance and repair ........ 308

**Index** ........ 311

# *Chapters one through thirteen*

# *chapter one*

# *Regulatory issues*

*"They who govern most make the least noise."*
—John Sidney

*"The essential problem is how to govern a large-scale world with small-scale local minds."*
—Sir Alfred Zimmern

Safety professionals should be aware of their rights and responsibilities when addressing the Occupational Safety and Health Administration (OSHA) or other governmental agencies before and after a disaster situation. Additionally, it is important that the safety professional be aware of the underlying purpose of the investigation and the potential civil and criminal penalties that can be assessed against the organization, as well as the safety professional or other members of the management team individually.

To begin with, the OSH Act covers virtually every American workplace that employs one or more employees and engages in a business that in any way affects interstate commerce.[1] The OSH Act covers employment in every state, the District of Columbia, Puerto Rico, Guam, the Virgin Islands, American Samoa, and the Trust Territory of the Pacific Islands.[2] The OSH Act does not, however, cover employees in situations where other state or federal agencies have jurisdiction that requires the agencies to prescribe or enforce their own safety and health regulations.[3] Additionally, the OSH Act exempts residential owners who employ people for ordinary domestic tasks, such as cooking, cleaning, and child care.[4] It also does not cover federal,[5] state, and local governments[6] or Native American reservations.[7]

The OSH Act does require every employer engaged in interstate commerce to furnish employees "a place of employment ... free from recognized hazards that are causing, or are likely to cause, death or serious harm."[8] To help employers create and maintain safe working environments and to enforce laws and regulations that ensure safe and healthful work environments, Congress provided for the creation of the Occupational Safety and Health Administration to be a new agency under the direction of the Department of Labor. Today, OSHA is one of the most widely known and powerful enforcement agencies. It has been granted broad regulatory powers to promulgate regulations and standards, investigate and inspect, issue citations, and propose penalties for safety violations in the workplace.

The OSH Act also established an independent agency to review OSHA citations and decisions, the Occupational Safety and Health Review Commission (OSHRC). The OSHRC is a quasi-judicial and independent administrative agency composed of three commissioners appointed by the president who serve staggered six-year terms. The OSHRC has the power to issue orders; uphold, vacate, or modify OSHA citations and penalties; and direct other appropriate relief and penalties.

The educational arm of the OSH Act is the National Institute for Occupational Safety and Health (NIOSH), which was created as a specialized educational agency of the existing National Institutes of Health. NIOSH conducts occupational safety and health research and develops criteria for new OSHA standards. NIOSH can conduct workplace inspections, issue subpoenas, and question employees and employers but it does not have the power to issue citations or penalties.

As permitted under the OSH Act, OSHA encourages individual states to take responsibility for OSHA administration and enforcement within their own respective boundaries. Each state possesses the ability to request and be granted the right to adopt state safety and health regulations and enforcement mechanisms.[9] For a state plan to be placed into effect, the state must first develop and submit its proposed program to the Secretary of Labor for review and approval. The Secretary must certify that the state plan's standards are "at least as effective" as the federal standards and that the state will devote adequate resources to administering and enforcing standards.[10]

In most state plans, the state agency has developed more stringent safety and health standards than OSHA[11] and has usually developed more stringent enforcement schemes.[12] The Secretary of Labor has no statutory authority to reject a state plan if the proposed standards or enforcement scheme are more strict than the OSHA standards but can reject the state plan if the standards are below the minimum limits set under OSHA standards.[13] These states are known as "state plan" states and territories.[14] (As of 1991, there were 21 states and two territories with approved and functional state plan programs.[15]) Employers in state plan states and territories must comply with their particular state's regulations; federal OSHA plays virtually no role in direct enforcement.

The Occupational Safety and Health Administration does, however, possess an approval and oversight role in regard to state plan programs. OSHA must approve all state plan proposals prior to enactment, and they maintain oversight authority to "pull the ticket" of any or all state plan programs at any time they are not achieving the identified prerequisites. Enforcement of this oversight authority was recently observed following a fire resulting in several workplace fatalities at the Imperial Foods facility in Hamlet, NC. Following this incident, federal OSHA assumed jurisdiction and control over the state plan program in North Carolina and made significant modifications to this program before returning the program to state control.

The OSH Act requires that a covered employer comply with specific occupational safety and health standards and all rules, regulations, and orders issued pursuant to the OSH Act that apply to the workplace.[16] The OSH Act also requires that all standards be based on research, demonstration, experimentation, or other appropriate information.[17] The Secretary of Labor is authorized under the Act to "promulgate, modify, or revoke any occupational safety and health standard,"[18] and the OSH Act describes the procedures that the Secretary must follow when establishing new occupational safety and health standards.[19]

The OSH Act authorizes three ways to promulgate new standards. From 1970 to 1973, the Secretary of Labor was authorized in Section 6(a) of the Act[20] to adopt national consensus standards and establish federal safety and health standards without following lengthy rulemaking procedures. Many of the early OSHA standards were adapted mainly from other areas of regulation, such as the National Electric Code and American National Standards Institute (ANSI) guidelines; however, this promulgation method is no longer in effect.

The usual method of issuing, modifying, or revoking a new or existing OSHA standard is set out in Section 6(b) of the OSH Act and is known as informal rulemaking. It requires notice to interested parties, through subscription to the *Federal Register*, of the proposed regulation and standard and provides an opportunity for comment in a nonadversarial administrative hearing.[21] The proposed standard can also be advertised through magazine articles and other publications, thus informing interested parties of the proposed standard and regulation. This method differs from the requirements of most other administrative agencies that follow the Administrative Procedure Act[22] in that the OSH Act provides interested persons an opportunity to request a public hearing with oral testimony. It also requires the Secretary of Labor to publish in the *Federal Register* a notice of the time and place of such hearings.

Although not required under the OSH Act, the Secretary of Labor has directed, by regulation, that OSHA follow a more rigorous procedure for comment and hearing than other administrative agencies.[23] Upon notice and request for a hearing, OSHA must provide a hearing examiner in order to listen to any oral testimony offered. All oral testimony is preserved in a verbatim transcript. Interested persons are provided an opportunity to cross-examine OSHA representatives or others on critical issues. The Secretary must state the reasons for the action to be taken on the proposed standard, and the statement must be supported by substantial evidence in the record as a whole.

The Secretary of Labor has the authority not to permit oral hearings and to call for written comment only. Within 60 days after the period for written comment or oral hearings has expired, the Secretary must decide whether to adopt, modify, or revoke the standard in question. The Secretary can also decide not to adopt a new standard. The Secretary must then publish a statement of the reasons for any decision in the *Federal Register.* OSHA regulations further mandate that the Secretary provide a supplemental statement of significant issues in the decision. Safety and health professionals should be aware that the standard as adopted and published in the *Federal Register* may be different from the proposed standard. The Secretary is not required to reopen hearings when the adopted standard is a "logical outgrowth" of the proposed standard.[24]

The final method for promulgating new standards, and the one most infrequently used, is the emergency temporary standard permitted under Section 6(c).[25] The Secretary of Labor may establish a standard immediately if it is determined that employees are subject to grave danger from exposure to substances or agents known to be toxic or physically harmful and that an emergency standard would protect the employees from the danger. An emergency temporary standard becomes effective on publication in the *Federal Register* and may remain in effect for six months. During this six-month period, the Secretary must adopt a new permanent standard or abandon the emergency standard.

Only the Secretary of Labor can establish new OSHA standards. Recommendations or requests for an OSHA standard can come from any interested person or organization, including employees, employers, labor unions, environmental groups, and others.[26] When the Secretary receives a petition to adopt a new standard or to modify or revoke an existing standard, he or she usually forwards the request to NIOSH and the National Advisory Committee on Occupational Safety and Health (NACOSH)[27] or the Secretary may use a private organization such as the American National Standards Institute (ANSI) for advice and review.

The OSH Act requires that an employer must maintain a place of employment free from recognized hazards that are causing or are likely to cause death or serious physical harm, even if there is no specific OSHA standard addressing the circumstances. Under Section 5(a)(1), known as the "general duty clause," an employer may be cited for a violation of the OSH Act if the condition causes harm or is likely to cause harm to employees, even if OSHA has not promulgated a standard specifically addressing the particular hazard. The general duty clause is a catch-all standard encompassing all potential hazards that have not been specifically addressed in the OSHA standards. For example, if a company is cited for an ergonomic hazard and there is no ergonomic standard to apply, the hazard will be cited under the general duty clause.

The OSH Act provides for a wide range of penalties, from a simple notice with no fine to criminal prosecution. The Omnibus Budget Reconciliation Act of 1990 multiplied maximum penalties sevenfold. Violations are categorized and penalties may be assessed as outlined in Table 1.

**Table 1** Violation and Penalty Schedule

| Penalty | Old Penalty Schedule (dollars) | New (1990) Penalty Schedule (dollars) |
|---|---|---|
| *De minimis* notice | 0 | 0 |
| Nonserious | 0–1000 | 0–7000 |
| Serious | 0–1000 | 0–7000 |
| Repeat | 0–10,000 | 0–70,000 |
| Willful | 0–10,000 | 25,000 minimum, 70,000 maximum |
| Failure to abate notice | 0–1000 per day | 0–7000 per day |
| New posting penalty | | 0–7000 |

Each alleged violation is categorized and the appropriate fine issued by the OSHA area director. It should be noted that each citation is separate and may carry with it a monetary fine. The gravity of the violation is the primary factor in determining penalties.[28] In assessing the gravity of a violation, the compliance officer or area director must consider: (1) the severity of the injury or illness that could result, and (2) the probability that an injury or illness could occur as a result of the violation.[29] Specific penalty assessment tables assist the area director or compliance officer in determining the appropriate fine for the violation.[30]

After selecting the appropriate penalty table, the area director or other official determines the degree of probability that the injury or illness will occur by considering:[31]

1. The number of employees exposed
2. The frequency and duration of the exposure
3. The proximity of employees to the point of danger
4. Factors such as the speed of the operation that require work under stress
5. Other factors that might significantly affect the degree of probability of an accident

OSHA has defined a serious violation as "an infraction in which there is a substantial probability that death or serious harm could result ... unless the employer did not or could not with the exercise of reasonable diligence, know of the presence of the violation."[32] Section 17(b) of the OSH Act requires that a penalty of up to $7000 be assessed for every serious violation cited by the compliance officer.[33] In assembly-line enterprises and manufacturing facilities with duplicate operations, if one process is cited as possessing a serious violation, it is possible that each of the duplicate processes or machines may be cited for the same violation. Thus, if a serious violation is found in one machine and there are many other identical machines in the enterprise, a very large monetary fine for a single serious violation is possible.[34]

Currently, the greatest monetary liabilities are for "repeat violations," "willful violations," and "failure to abate" cited violations. A *repeat* violation is a second citation for a violation that was cited previously by a compliance officer. OSHA maintains records of all violations and must check for repeat violations after each inspection. A *willful* violation is the employer's purposeful or negligent failure to correct a known deficiency. This type of violation, in addition to carrying a large monetary fine, exposes the employer to a charge of an "egregious" violation and the potential for criminal sanctions under the OSH Act or state criminal statutes if an employee is injured or killed as a direct result of the willful violation. *Failure to abate* a cited violation has the greatest cumulative monetary liability of all. OSHA may assess a penalty of up to $1000 per day per violation for each day in which a cited violation is not brought into compliance.

In assessing monetary penalties, the area or regional director must consider the good faith of the employer, the gravity of the violation, the employer's past history of compliance, and the size of the employer. Joseph Dear, the Assistant Secretary of Labor, recently stated that OSHA will start using its egregious case policy, which has seldom been invoked in recent years.[35] Under the egregious violation policy, when violations are determined to be conspicuous, penalties are cited for each violation, rather than combining the violations into a single, smaller penalty.

In addition to the potential civil or monetary penalties that could be assessed, OSHA regulations may be used as evidence in negligence, product liability, workers' compensation, and other actions involving employee safety and health issues.[36] OSHA standards and regulations are the baseline requirements for safety and health that must be met, not only to achieve compliance with the OSHA regulations, but also to safeguard an organization against other potential civil actions.

The OSHA monetary penalty structure is classified according to the type and gravity of the particular violation. Violations of OSHA standards or the general duty clause are categorized as "*de minimis*,"[37] "other" (nonserious),[38] "serious,"[39] "repeat,"[40] and "willful."[41] See Table 1 for penalty schedules. Monetary penalties assessed by the Secretary vary according to the degree of the violation. Penalties range from no monetary penalty to 10 times the imposed penalty for repeat or willful violations.[42] Additionally, the Secretary may refer willful violations to the U.S. Department of Justice for imposition of criminal sanctions.[43]

## De minimis *violations*

When a violation of an OSHA standard does not immediately or directly relate to safety or health, OSHA either does not issue a citation or issues a "*de minimis*" citation. Section 9 of the OSH Act provides that "[the] Secretary may prescribe procedures for the issuance of a notice in lieu of a citation with respect to *de minimis* violations which have no direct or immediate relationship to safety or health."[44] A *de minimis* notice does not constitute a citation and no fine is imposed. Additionally, there usually is no abatement period and thus there can be no violation for failure to abate.

The *OSHA Compliance Field Operations Manual* (*OSHA Manual*)[45] provides two examples of when *de minimis* notices are generally appropriate: (1) "in situations involving standards containing physical specificity wherein a slight deviation would not have an immediate or direct relationship to safety or health,"[46] and (2) "where the height of letters on an exit sign is not in strict conformity with the size requirements of the standard."[47]

The Occupational Safety and Health Administration has found *de minimis* violations in cases where employees, as well as the safety records, are persuasive in exemplifying that no injuries or lost time have been incurred.[48] Additionally, in order for OSHA to conserve valuable resources to produce a greater impact on safety and health in the workplace, it is highly likely that the Secretary will encourage use of the *de minimis* notice in marginal cases and even in other situations where the possibility of injury is remote and potential injuries would be minor.

## *Other or nonserious violations*

"Other" or nonserious violations are issued where a violation could lead to an accident or occupational illness, but the probability that it would cause death or serious physical harm is minimal. Such a violation, however, does possess a direct or immediate relationship to the safety and health of workers.[49] Potential penalties for this type of violation range from no fine up to $7000 per violation.[50]

In distinguishing between a serious and a nonserious violation, the OSHRC has stated that "a nonserious violation is one in which there is a direct and immediate relationship between the violative condition and occupational safety and health but no such relationship that a result of injury or illness is death or serious physical harm."[51] The *OSHA Manual* provides guidance and examples for issuing nonserious violations. It states that:

> "...an example of nonserious violation is the lack of guardrail at a height from which a fall would more probably result in only a mild sprain or cut or abrasion; i.e., something less than serious harm.[52]
>
> "A citation for serious violation may be issued or a group of individual violations [which] taken by themselves would be nonserious, but together would be serious in the sense that in combination they present a substantial probability of injury resulting in death or serious physical harm to employees.[53]

> "A number of nonserious violations [which] are present in the same piece of equipment which, considered in relation to each other, affect the overall gravity of possible injury resulting from an accident involving the combined violations ... may be grouped in a manner similar to that indicated in the preceding paragraph, although the resulting citation will be for a nonserious violation."[54]

The difference between a serious and a nonserious violation hinges on subjectively determining the probability of injury or illness that might result from the violation. Administrative decisions have usually turned on the particular facts of the situation. The OSHRC has reduced serious citations to nonserious violations when the employer was able to show that the probability of an accident, and the probability of a serious injury or death, was minimal.[55]

## *Serious violations*

Section 17(k) of the OSH Act defines a serious violation as one where:

> "...there is a substantial probability that death or serious physical harm could result from a condition which exists, or from one or more practices, means, methods, operations or processes which have been adopted or are in use, in such place of employment unless the employer did not, and could not with exercise of reasonable diligence, know of the presence of the violation."[56]

Section 17(b) of the Act provides that a civil penalty of up to $7000 must be assessed for serious violations, whereas for nonserious violations civil penalties may be assessed.[57] The amount of the penalty is determined by considering: (1) the gravity of the violation, (2) the size of the employer, (3) the good faith of the employer, and (4) the employer's history of previous violations.[58]

To prove that a violation is within the serious category, OSHA must only show a substantial probability that a foreseeable accident would result in serious physical harm or death. Thus, contrary to common belief, OSHA does not need to show that a violation would create a high probability that an accident would result. Because substantial physical harm is the distinguishing factor between a serious and a nonserious violation, OSHA has defined "serious physical harm" as "permanent, prolonged, or temporary impairment of the body in which part of the body is made functionally useless or is substantially reduced in efficiency on or off the job." Additionally, an occupational illness is defined as "illness that could shorten life or significantly reduce physical or mental efficiency by inhibiting the normal function of a part of the body."[59]

After determining that a hazardous condition exists and that employees are exposed or potentially exposed to the hazard, the *OSHA Manual* instructs compliance officers to use a four-step approach to determine whether the violation is serious:[60]

1. Determine the type of accident or health hazard exposure that the violated standard is designed to prevent in relation to the hazardous condition identified.
2. Determine the type of injury or illness which it is reasonably predictable could result from the type of accident or health hazard exposure identified in step 1.
3. Determine that the type of injury or illness identified in step 2 includes death or a form of serious physical harm.
4. Determine that the employer knew or with the exercise of reasonable diligence could have known of the presence of the hazardous condition.

The *OSHA Manual* provides examples of serious injuries, including amputations, fractures, deep cuts involving extensive suturing, disabling burns, and concussions. Examples of serious illnesses include cancer, silicosis, asbestosis, poisoning, and hearing and visual impairment.[61]

Safety professionals should be aware that OSHA is not required to show that the employer actually knew that the cited condition violated safety or health standards. The employer can be charged with constructive knowledge of the OSHA standards. OSHA also does not have to show that the employer could reasonably foresee that an accident would happen, although it does have the burden of proving that the possibility of an accident was not totally unforeseeable. OSHA does need to prove, however, that the employer knew or should have known of the hazardous condition and that it knew there was a substantial likelihood that serious harm or death would result from an accident.[62] If the Secretary cannot prove that the cited violation meets the criteria for a serious violation, the violation may be cited in one of the lesser categories.

## *Willful violations*

The most severe monetary penalties under the OSHA penalty structure are for willful violations. A "willful" violation can result in penalties of up to $70,000 per violation, with a minimum required penalty of $5000. Although the term "willful" is not defined in OSHA regulations, courts generally have defined a willful violation as "an act voluntarily with either an intentional disregard of, or plain indifference to, the Act's requirements."[63] Further, the OSHRC defines a willful violation as "action taken knowledgeably by one subject to the statutory provisions of the OSH Act in disregard of the action's legality. No showing of malicious intent is necessary. A conscious, intentional, deliberate, voluntary decision is properly described as willful."[64]

There is little distinction between civil and criminal willful violations other than the due process requirements for a criminal violation and the fact that a violation of the general duty clause cannot be used as the basis for a criminal willful violation. The distinction is usually based on the factual circumstances and the fact that a criminal willful violation results from a willful violation which caused an employee death.

According to the *OSHA Manual*, the compliance officer "can assume that an employer has knowledge of any OSHA violation condition of which its supervisor has knowledge; he can also presume that, if the compliance officer was able to discover a violative condition, the employer could have discovered the same condition through the exercise of reasonable diligence."[65]

Courts and the OSHRC have agreed on three basic elements of proof that OSHA must show for a willful violation. OSHA must show that the employer (1) knew or should have known that a violation existed, (2) voluntarily chose not to comply with the OSH Act to remove the violative condition, and (3) made the choice not to comply, with intentional disregard of the OSH Act's requirements or plain indifference to them properly characterized as reckless. Courts and the OSHRC have affirmed findings of willful violations in many circumstances, ranging from deliberate disregard of known safety requirements[66] through fall protection equipment not being provided.[67] Other examples of willful violations include cases where safety equipment was ordered but employees were permitted to continue work until the equipment arrived,[68] inexperienced and untrained employees were permitted to perform a hazardous job,[69] and an employer failed to correct a situation that had been previously cited as a violation.

## *Repeat and failure to abate violations*

"Repeat" and "failure to abate" violations are often quite similar and confusing to human resources (HR) professionals. When, upon reinspection by OSHA, a violation of a previously cited standard is found but the violation does not involve the same machinery, equipment, process, or location, this would constitute a repeat violation. If, upon reinspection by OSHA, a violation of a previously cited standard is found but evidence indicates that the violation continued uncorrected since the original inspection, this would constitute a failure to abate violation.[70]

The most costly civil penalty under the OSH Act is for repeat violations. The OSH Act authorizes a penalty of up to $70,000 per violation but permits a maximum penalty of 10 times the maximum authorized for the first instance of the violation. Repeat violations can also be grouped within the willful category (i.e., a willful repeat violation) to acquire maximum civil penalties.

In certain cases where an employer has more than one fixed establishment and citations have been issued, the *OSHA Manual* states:[71]

> "...for the purpose for considering whether a violation is repeated, citations issued to employers having fixed establishments (e.g., factories, terminals, stores) will be limited to the cited establishment. ...For employers engaged in businesses having no fixed establishments, repeated violations will be alleged based upon prior violations occurring anywhere within the same Area Office Jurisdiction."

When a previous citation has been contested but a final OSHRC order has not yet been received, a second violation is usually cited as a repeat violation. The *OSHA Manual* instructs the compliance officer to notify the assistant regional director and to indicate on the citation that the violation is contested.[72] If the first citation never becomes a final OSHRC order (i.e., the citation is vacated or otherwise dismissed), the second citation for the repeat violation will be removed automatically.[73]

As noted previously, a failure to abate violation occurs when, upon reinspection, the compliance officer finds that the employer has failed to take necessary corrective action and thus the violation continues uncorrected. The penalty for a failure to abate violation can be up to $7000 per day to a maximum of $70,000. HR professionals should also be aware that citations for repeat violations, failure to abate violations, or willful repeat violations can be issued for violations of the general duty clause. The *OSHA Manual* instructs compliance officers that citations under the general duty clause are restricted to serious violations or to willful or repeat violations that are of a serious nature.[74]

## *Failure to post violation notices*

A new penalty category, the failure to post violation notices carries a penalty of up to $7000 for each violation. A failure to post violation occurs when an employer fails to post notices required by the OSHA standards, including the OSHA poster, a copy of the year-end summary of the OSHA 200 form, a copy of OSHA citations when received, and copies of other pleadings and notices.

## *Criminal liability and penalties*

The OSH Act provides for criminal penalties in four circumstances.[75] In the first, anyone inside or outside of the Department of Labor or OSHA who gives advance notice of an inspection, without authority from the Secretary, may be fined up to $1000, imprisoned for up to six months, or both. Second, any employer or person who intentionally falsifies statements or OSHA records that must be prepared, maintained, or submitted under the OSH Act may, if found guilty, be fined up to $10,000, imprisoned for up to six months, or both. Third, any person responsible for a violation of an OSHA standard, rule, order, or regulation who causes the death of an employee may, upon conviction, be fined up to $10,000, imprisoned for up to six months, or both. If convicted for a second violation, punishment may be a fine of up to $20,000, imprisonment for up to one year, or both.[76] Finally, if an individual is convicted of forcibly resisting or assaulting a compliance officer or other Department of Labor personnel, a fine of $5000, three years in prison, or both, can be imposed. Any person convicted of killing a compliance officer or other OSHA or Department of Labor personnel acting in his or her official capacity may be sentenced to prison for any term of years or life.

The Occupational Safety and Health Administration does not have authority to impose criminal penalties directly; instead, it refers cases for possible criminal prosecution to the U.S. Department of Justice. Criminal penalties must be based on violation of a specific OSHA standard; they may not be based on a violation of the general duty clause. Criminal prosecutions are conducted like any other criminal trial, with the same rules of evidence, burden of proof, and rights of the accused. A corporation may be criminally liable for the acts of its agents or employees.[77] The statute of limitations for possible criminal violations of the OSH Act, as for other federal noncapital crimes, is five years.[78] Under federal criminal law, criminal charges may range from murder to manslaughter to conspiracy. Several charges may be brought against an employer for various separate violations under one federal indictment.

The OSH Act provides for criminal penalties of up to $10,000 and/or imprisonment for up to six months. A repeated willful violation causing an employee death can double the criminal sanction to a maximum of $20,000 and/or one year of imprisonment. Given the increased use of criminal sanctions by OSHA in recent years, personnel managers should advise their employers of the potential for these sanctions being used when the safety and health of employees are disregarded or put on the back burner.

Criminal liability for a willful OSHA violation can attach to an individual or a corporation. In addition, corporations may be held criminally liable for the actions of their agents or officials.[79] Safety professionals and other corporate officials may also be subject to criminal liability under a theory of aiding and abetting the criminal violation in their official capacity with the corporation.[80]

Safety professionals should also be aware that an employer could face two prosecutions for the same OSHA violation without the protection of double jeopardy. The OSHRC can bring an action for a civil willful violation using the monetary penalty structure described previously and the case then may be referred to the Justice Department for criminal prosecution of the same violation.[81]

Prosecution of willful criminal violations by the Justice Department has been rare in comparison to the number of inspections performed and violations cited by OSHA on a yearly basis. However, the use of criminal sanctions has increased substantially in the last few years. With adverse publicity being generated as a result of workplace accidents and deaths[82] and Congress emphasizing reform, a decrease in criminal prosecutions is unlikely.

The law regarding criminal prosecution of willful OSH Act violations is still emerging. Although few cases have actually gone to trial, in most situations the mere threat of criminal prosecution has encouraged employers to settle cases with assurances that criminal prosecution would be dismissed. Many state plan states are using criminal sanctions permitted under their state OSH regulations more frequently.[83] State prosecutors have also allowed use of state criminal codes for workplace deaths.[84]

After a disaster situation, especially if a fatality is involved, safety professionals should exercise extreme caution. The potential for criminal sanctions and criminal prosecution is substantial if a willful violation of a specific OSHA standard is directly involved in the death. The OSHA investigation may be conducted from a criminal perspective in order to gather and secure the appropriate evidence to later pursue criminal sanctions.[85] A prudent safety professional facing a workplace fatality investigation should address the OSHA investigation with legal counsel present and reserve all rights guaranteed under the U.S. Constitution.[86] Obviously, under no circumstances should a safety professional condone or attempt to conceal facts or evidence, which consists of a cover-up.

The Occupational Safety and Health Administration performs all enforcement functions under the OSH Act. Under Section 8(a) of the Act, OSHA compliance officers have the right to enter any workplace of a covered employer without delay, to inspect and investigate a workplace during regular hours and at other reasonable times, and to obtain an inspection warrant if access to a facility or operation is denied.[87] Upon arrival at an inspection site, the compliance officer is required to present his credentials to the owner or designated representative of the employer before starting the inspection. The employer representative and an employee and/or union representative may accompany the compliance officer on the inspection. Compliance officers can question the employer and employees and inspect required records, such as the OSHA Form 200, which records injuries and illnesses.[88] Most compliance officers cannot issue on-the-spot citations; they only have authority to document potential hazards and report or confer with the OSHA area director before issuing a citation.

A compliance officer or any other employee of OSHA may not provide advance notice of the inspection under penalty of law.[89] The OSHA area director is, however, permitted to provide notice under the following circumstances:[90]

1. In cases of apparent imminent danger, to enable the employer to correct the danger as quickly as possible
2. When the inspection can most effectively be conducted after regular business hours or where special preparations are necessary
3. To ensure the presence of employee and employer representatives or appropriate personnel needed to aid in inspections
4. When the area director determines that advance notice would enhance the probability of an effective and thorough inspection

Compliance officers can also take environmental samples and obtain photographs related to the inspection. Additionally, compliance officers, can use other "reasonable investigative techniques," including personal sampling equipment, dosimeters, air-sampling badges, and other equipment.[91] Compliance officers must, however, take reasonable precautions when using photographic or sampling equipment to avoid creating hazardous conditions (i.e., a spark-producing camera flash in a flammable area) or disclosing a trade secret.[92]

An OSHA inspection has four basic components: (1) the opening conference, (2) the walk-through inspection, (3) the closing conference, and (4) the issuance of citations, if necessary. In the opening conference, the compliance officer may explain the purpose and type of inspection to be conducted, request records to be evaluated, question the employer, ask for appropriate representatives to accompany him during the walk-through inspection, and ask additional questions or request more information. The compliance officer may, but is not required to, provide the employer with copies of the applicable laws and regulations governing procedures and health and safety standards. The opening conference is usually brief and informal, its primary purpose being to establish the scope and purpose of the walk-through inspection.

After the opening conference and review of appropriate records, the compliance officer, usually accompanied by a representative of the employer and a representative of the employees, conducts a physical inspection of the facility or worksite.[93] The general purpose of this walk-through inspection is to determine whether the facility or worksite complies with OSHA standards. The compliance officer must identify potential safety and health hazards in the workplace, if any, and document them to support issuance of citations.[94]

The compliance officer uses various forms to document potential safety and health hazards observed during the inspection. The most commonly used form is the OSHA-1 Inspection Report, in which the compliance officer records information gathered during the opening conference and walk-through inspection.[95]

Two additional forms are usually attached to the OSHA Inspection Report. The OSHA-1A form, known as the narrative, is used to record information gathered during the walk-through inspection; names and addresses of employees, management officials, and employee representatives accompanying the compliance officer on the inspection; and other information. A separate worksheet, known as OSHA-1B, is used by the compliance officer to document each condition that he believes could be an OSHA violation. One OSHA-1B worksheet is completed for each potential violation noted by the compliance officer.

When the walk-through inspection is completed, the compliance officer usually conducts an informal meeting with the employer or the employer's representative to "informally advise (the employer) of any apparent safety or health violations disclosed by the inspection.[96] The compliance officer informs the employer of the potential hazards observed and indicates the applicable section of the standards allegedly violated, advises that citations may be issued, and informs the employer or representative of the appeal process and rights.[97] Additionally, the compliance officer advises the employer that the OSH Act prohibits discrimination against employees or others for exercising their rights.[98]

In an unusual situation, the compliance officer may issue a citation(s) on the spot. When this occurs, the compliance officer informs the employer of the abatement period, in addition to the other information provided at the closing conference. In most circumstances, the compliance officer will leave the workplace and file a report with the area director who has authority, through the Secretary of Labor, to decide whether a citation should be issued, compute any penalties to be assessed, and set the abatement date for each alleged violation. The area director, under authority from the Secretary, must issue the citation with "reasonable promptness."[99] Citations must be issued in writing and must describe with particularity the violation alleged, including the relevant standard and regulation. There is a six-month statute of limitations, and the citation must be issued or vacated within this time period. OSHA must serve notice of any citation and proposed penalty by certified mail, unless there is personal service, to an agent or officer of the employer.[100]

After the citation and notice of proposed penalty are issued, but before the notice of contest by the employer is filed, the employer may request an informal conference with the OSHA area director. The general purpose of the informal conference is to clarify the basis for the citation, modify abatement dates or proposed penalties, seek withdrawal of a cited item, or otherwise attempt to settle the case. This conference, as its name implies, is an informal meeting between the employer and OSHA. Employee representatives must have an opportunity to participate if they so request. Safety professionals should note that the request for an informal conference does not "stay" (delay) the 15-working-day period to file a notice of contest to challenge the citation.[101]

Under the OSH Act, an employer, employee, or authorized employee representative (including a labor organization) is given 15 working days from when the citation is issued to file a "notice of contest." If a notice of contest is not filed within 15 working days, the citation and proposed penalty become a final order of the Occupational Safety and Health Review Commission (OSHRC) and are not subject to review by any court or agency. If a timely notice of contest is filed in good faith, the abatement requirement is tolled (temporarily suspended or delayed) and a hearing is scheduled. The employer also has the right to file a petition for modification of the abatement period (PMA) if the employer is unable to comply with the abatement period provided in the citation. If OSHA contests the PMA, a hearing is scheduled to determine whether the abatement requirements should be modified.

When the notice of contest by the employer is filed, the Secretary must immediately forward the notice to the OSHRC, which then schedules a hearing before its administrative law judge (ALJ). The Secretary of Labor is labeled the "complainant," and the employer the "respondent." The ALJ may affirm, modify, or vacate the citation, any penalties, or the abatement date. Either party can appeal the administrative law judge's decision by filing a petition for discretionary review (PDR). Additionally, any member of the OSHRC may "direct review" of any decision by an ALJ, in whole or in part, without a PDR. If a PDR is not filed and no member of the OSHRC directs a review, the decision of the ALJ becomes final in 30 days. Any party may appeal a final order of the OSHRC by filing a petition for review in the U.S. Court of Appeals for the circuit in which the violation is alleged to have occurred or in the U.S. Court of Appeals for the District of Columbia Circuit. This petition for review must be filed within 60 days from the date of the OSHRC's final order.

## *OSHA inspection checklist*

The following is a recommended checklist for safety professionals in order to prepare for an OSHA inspection in general or after a disaster situation:[102]

1. Assemble a team from the management group and identify specific responsibilities in writing for each team member. The team members should be given appropriate training and education and should include, but not be limited to:
   a. OSHA inspection team coordinator
   b. Document-control individual
   c. Individuals to accompany the OSHA inspector
   d. Media coordinator
   e. Accident investigation team leader (where applicable)
   f. Notification person
   g. Legal advisor (where applicable)
   h. Law enforcement coordinator (where applicable)
   i. Photographer
   j. Industrial hygienist
2. Decide on and develop a company policy and procedures to provide guidance to the OSHA inspection team.
3. Prepare an OSHA inspection kit, including all equipment necessary to properly document all phases of the inspection. The kit should include equipment such as a camera (with extra film and batteries), a tape player (with extra batteries), a video camera, pads, pens, and other appropriate testing and sampling equipment (e.g., noise-level meter, air-sampling kit, etc.).
4. Prepare basic forms to be used by the inspection team members during and following the inspection.
5. When notified that an OSHA inspector has arrived, assemble the team members along with the inspection kit.
6. Identify the inspector. Check his credentials, and determine the reason for and type of inspection to be conducted.

7. Confirm the reason for the inspection with the inspector (targeted, routine inspection, accident, or in response to a complaint).
   a. For a random or target inspection:
      - Did the inspector check the OSHA 200 Form?
      - Was a warrant required?
   b. For an employee complaint inspection:
      - Did inspector have a copy of the complaint? If so, obtain a copy.
      - Do allegations in the complaint describe an OSHA violation?
      - Was a warrant required?
      - Was the inspection protested in writing?
   c. For an accident investigation inspection:
      - How was OSHA notified of the accident?
      - Was a warrant required?
      - Was the inspection limited to the accident location?
   d. If a warrant is presented:
      - Were the terms of the warrant reviewed by local counsel?
      - Did the inspector follow the terms of the warrant?
      - Was a copy of the warrant acquired?
      - Was the inspection protested in writing?
8. The opening conference:
   a. Who was present?
   b. What was said?
   c. Was the conference taped or otherwise documented?
9. Records:
   a. What records were requested by the inspector?
   b. Did the document-control coordinator number the photocopies of the documents provided to the inspector?
   c. Did the document control coordinator maintain a list of all photocopies provided to the inspector?
10. Facility inspection:
   a. What areas of the facility were inspected?
   b. What equipment was inspected?
   c. Which employees were interviewed?
   d. Who was the employee or union representative present during the inspection?
   e. Were all the remarks made by the inspector documented?
   f. Did the inspector take photographs?
   g. Did a team member take similar photographs?[102]

In a situation where local or state prosecutors (i.e., district attorney, or DA) respond to a fatality or disaster situation, safety professionals should exercise extreme caution and acquire legal representation prior to providing any information or documents. The prosecutor is usually investigating from a criminal perspective and would utilize the state criminal code as the basis for any charges. Safety professionals should preserve all constitutional rights, including the right to remain silent and the right to counsel in these situations.

As noted in the Miranda warnings provided by law enforcement after an arrest, "You have the right to remain silent; everything you say can and will be used against you in a court of law; you have the right to counsel…"

In summation, being prepared and "knowing the rules" can assist safety professionals in protecting their organization's assets during any type of governmental regulatory inspection. Although we have used OSHA as an example due to the fact that most safety professionals interact with OSHA more than other governmental agencies, safety professionals with responsibilities in human resources, environmental, and other areas should have an in-depth familiarity with the regulatory scheme and rules of the specific governmental entity. Preparation and knowledge will permit smooth sailing during any inspection!

## Endnotes

1. 29 C.F.R. §1975.3(6).
2. *Id.* at §652(7).
3. See, e.g., Atomic Energy Act of 1954, 42 U.S.C. §2021.
4. 29 C.F.R. §1975(6).
5. 29 U.S.C.A. §652(5) (no coverage under OSH Act, when U.S. government acts as employer).
6. *Id.*
7. See, e.g., *Navajo Forest Prods. Indus.*, 8 O.S.H. Cases 2694 (OSH Rev. Comm'n 1980), aff'd, 692 F.2d 709, 10 O.S.H. Cases 2159.
8. 29 U.S.C.A. §654(a)(1).
9. In Section 18(b), the OSH Act provides that any state "which, at any time, desires to assume responsibility for development and the enforcement therein of occupational safety and health standards relating to any … issue with respect to which a federal standard has been promulgated … shall submit a state plan for the development of such standards and their enforcement."
10. *Id.* at §667(c). After an initial evaluation period of a least three years during which OSHA retains concurrent authority, a state with an approved plan gains exclusive authority over standard settings, inspection procedures, and enforcement of health and safety issues covered under the state plan. See also *Noonan v. Texaco,* 713 P.2d 160 (Wyo. 1986); Plans for the Development and Enforcement of State Standards, 29 C.F.R. §667(f) (1982) and §1902.42(c)(1986). Although the state plan is implemented by the individual state, OSHA continues to monitor the program and may revoke the state authority if the state does not fulfill the conditions and assurances contained within the proposed plan.
11. Some states incorporate federal OSHA standards into their plans and add only a few of their own standards as a supplement. Other states, such as Michigan and California, have added a substantial number of separate and independently promulgated standards. See, generally, Employee Safety and Health Guide (CCH) §§5000–5840 (1987) (compiling all state plans). Some states also add their own penalty structures. For example, under Arizona's plan, employers may be fined up to $150,000 and sentenced to one and one-half years in prison for knowing violations of state standards that cause death to an employee and may also have to pay $25,000 in compensation to the victim's family. If the employer is a corporation,

the maximum fine is $1 million. See *Ariz. Rev. Stat. Ann.* §§13-701, 13-801, 23-4128, 23-418.01, 13-803 (Supp. 1986).

[12] For example, under Kentucky's state plan regulations for controlling hazardous energy (i.e., lockout/tagout), locks would be required rather than locks or tags being optional, as under the federal standard. Lockout/tagout is discussed in more detail in Chapter 2.

[13] 29 U.S.C. §667.

[14] 29 U.S.C.A. §667; 29 C.F.R. §1902.

[15] The states and territories operating their own OSHA programs are Alaska, Arizona, California, Hawaii, Indiana, Iowa, Kentucky, Maryland, Michigan, Minnesota, Nevada, New Mexico, North Carolina; partial federal OSHA enforcement, Oregon, Puerto Rico, South Carolina, Tennessee, Utah, Vermont, Virgin Islands, Virginia, Washington, and Wyoming.

[16] 29 U.S.C. §655(b).

[17] 29 U.S.C.A. §655(b)(5).

[18] 29 U.S.C. 1910.

[19] 29 C.F.R. §1911.15. By regulation, the Secretary of Labor has prescribed more detailed procedures than the OSH Act specifies to ensure participation in the process of setting new standards.

[20] 29 U.S.C. §1910.

[21] 29 U.S.C. §655(b).

[22] 5 U.S.C. §553.

[23] 29 C.F.R. §1911.15.

[24] *Taylor Diving & Salvage Co. v. Department of Labor,* 599 F.2d 622 7 O.S.H. Cases 1507 (5th Cir. 1979).

[25] 29 U.S.C. §655(c).

[26] *Id.* at §655(b)(1).

[27] *Id.* at §656(a)(1). NACOSH was created by the OSH Act to "advise, consult with, and make recommendations … on matters relating to the administration of the Act." Normally, for new standards, the Secretary has established continuing committees and *ad hoc* committees to provide advice regarding particular problems or proposed standards.

[28] *OSHA Compliance Field Operations Manual* (*OSHA Manual*) at XI-C3c (Apr. 1977).

[29] *Id.*

[30] *Id.* at XI-C3c(2).

[31] *Id.* at (3)(a).

[32] 29 U.S.C. §666.

[33] *Id.* at §666(b).

[34] For example, if a company possesses 25 identical machines, and each of these machines is found to have the identical serious violation, this would theoretically constitute 25 violations, rather than one violation on 25 machines, and a possible monetary fine of $175,000, rather than a maximum of $7000.

[35] *Occupational Safety & Health Reporter,* V. 23, No. 32, Jan. 12, 1994.

[36] See *infra* at §1.140.

[37] 29 U.S.C.§§658(a), 666(c).

[38] *Id.* at §666(j).

[39] *Id.* at §666(c).

[40] *Id.* at §666 (a).

[41] *Id.*

[42] *Id.* at (b).

[43] *Id.* at (e).
[44] *Id.* at §658(a).
[45] *Supra* at note 62.
[46] *Id.* at VII-B3a.
[47] *Id.*
[48] *Hood Sailmakers,* 6 O.S.H. Cases 1207 (1977).
[49] *OSHA Manual, supra* at note 62, at VIII-B2a. The proper nomenclature for this type of violation is "other" or "other than serious." Many safety and health professionals classify this type of violation as nonserious for explanation and clarification purposes.
[50] A nonserious penalty is usually less than $100 per violation.
[51] *Crescent Wharf & Warehouse Co.,* 1 O.S.H. Cases 1219, 1222 (1973).
[52] *OSHA Manual, supra* at note 62, at VIII-B2a.
[53] *Id.* at B2b(1).
[54] *Id.* at (2).
[55] See *Secretary v. Diamond Indus.,* 4 O.S.H. Cases 1821 (1976); *Secretary v. Northwest Paving,* 2 O.S.H. Cases 3241 (1974); *Secretary v. Sky-Hy Erectors & Equip.,* 4 O.S.H. Cases 1442 (1976). But see *Shaw Constr. v. OSHRC,* 534 F.2d 1183, 4 O.S.H. Cases 1427 (5th Cir. 1976) (holding that serious citation was proper whenever accident was merely possible.)
[56] 29 U.S.C. §666(j).
[57] *Id.*
[58] 29 U.S.C. §666(i).
[59] *OSHA Manual, supra* at note 62, at IV-B-1(b)(3)(a),(c).
[60] *Id.* at VIII-B1b(2)(c). In determining whether a violation constitutes a serious violation, the compliance officer is functionally describing the *prima facie* case that the Secretary would be required to prove, i.e., (1) the causal link between the violation of the safety or health standard and the hazard, (2) reasonably predictable injury or illness that could result, (3) potential of serious physical harm or death, and (4) the employer's ability to foresee such harm by using reasonable diligence.
[61] *Id.* at VIII-B1c(3)a.
[62] *Id.* at (4). See also *Cam Indus.,* 1 O.S.H. Cases 1564 (1974); *Secretary v. Sun Outdoor Advertising,* 5 O.S.H. Cases 1159 (1977).
[63] *Cedar Constr. Co. v. OSHRC,* 587 F.2d 1303, 6 O.S.H. Cases 2010, 2011 (D.C. Cir. 1971). Moral turpitude or malicious intent are not necessary elements for a willful violation. *U.S. v. Dye Constr.,* 522 F.2d 777, 3 O.S.H. Cases 1337 (4th Cir. 1975); *Empire-Detroit Steel v. OSHRC,* 579 F.2d 378, 6 O.S.H. Cases 1693 (6th Cir. 1978).
[64] *P.A.F. Equip. Co.,* 7 O.S.H. Cases 1209 (1979).
[65] *OSHA Manual, supra* at note 62, at VIII-B1c(4).
[66] *Universal Auto Radiator Mfg. Co. v. Marshall,* 631 F.2d 20, 8 O.S.H. Cases 2026 (3d Cir. 1980).
[67] *Haven Steel Co. v. OSHRC,* 738 F.2d 397, 11 O.S.H. Cases 2057 (10th Cir. 1984).
[68] *Donovan v. Capital City Excavating Co.,* 712 F.2d 1008, 11 O.S.H. Cases 1581 (6th Cir. 1983).
[69] *Ensign-Bickford Co. v. OSHRC,* 717 F.2d 1419, 11 O.S.H. Cases 1657 (D.C. Cir. 1983).
[70] *OSHA Manual, supra* at note 62, at VIII-B5c.
[71] *Id.* at IV-B5(c)(1).
[72] *Id.* at VIII-B5d.
[73] *Id.*
[74] *Id.* at XI-C5c.

[75] 29 U.S.C. §666(e)–(g). See also *OSHA Manual, supra* note 62 at VI-B.
[76] A repeat criminal conviction for a willful violation causing an employee death doubles the possible criminal penalties.
[77] 29 C.F.R. §5.01(6).
[78] *U.S. v. Dye Const. Co.*, 510 F.2d 78, 2 O.S.H. Cases 1510 (10th Cir. 1975).
[79] *U.S. v. Crosby & Overton*, No. CR-74-1832-F (S.D. Cal. Feb. 24, 1975).
[80] 18 U.S.C. §2.
[81] These are uncharted waters. Employers may argue due process and double jeopardy, but OSHA may argue that it has authority to impose penalties in both contexts. There are currently no cases on this issue.
[82] Jefferson, Dying for work, *A.B.A.*, #J. 46 (Jan.), 1993.
[83] See, e.g., Levin, Crimes against employees: substantive criminal sanctions under the Occupational Safety and Health Act, *Am. Crim. L. Rev.*, 14, 98, 1977.
[84] See Chapter 5.
[85] See L.A. law: prosecuting workplace killers, *A.B.A.*, #J. 48. (Los Angeles prosecutor's "roll out" program could serve as model for OSHA.)
[86] See 29 C.F.R. §1903.8.
[87] See *infra* §§1.10 and 1.12.
[88] 29 C.F.R. §1903.8.
[89] 29 U.S.C. §17(f). The penalty for providing advance notice, upon conviction, is a fine of not more than $1000, imprisonment for not more than 6 months, or both.
[90] *Occupational Safety and Health Law*, 208–209 (1988).
[91] 29 C.F.R. §1903.7(b) (revised by 47 *Fed. Reg.* 5548 [1982].)
[92] See, e.g., 29 C.F.R. §1903.9. Under §15 of the OSH Act, all information gathered or revealed during an inspection or proceeding that may reveal a trade secret as specified under 18 U.S.C. §1905 must be considered confidential, and breach of that confidentiality is punishable by a fine of not more than $1000, imprisonment of not more than one year, or both, and removal from office or employment with OSHA.
[93] It is highly recommended by the authors that a company representative accompany the OSHA inspection during the walk-through inspection.
[94] *OSHA Manual, supra* note 62, at III-D8.
[95] The OSHA 1 Inspection Report Form includes the following:

- Establishment's name
- Inspection number
- Type of legal entity
- Type of business or plant
- Additional citations
- Names and addresses of all organized employee groups
- Authorized representative of employees
- Employee representative contacted
- Other persons contacted
- Coverage information (state of incorporation, type of goods or services in interstate commerce, etc.)
- Date and time of entry
- Date and time that the walk-through inspection began
- Date and time closing conference began
- Date and time of exit
- Whether a follow-up inspection is recommended
- Compliance officer's signature and date

- Names of other compliance officers,
- Evaluation of safety and health programs (checklist)
- Closing conference checklist
- Additional comments

[96] 29 C.F.R. §1903.7(e).
[97] *OSHA Manual, supra* at note 62, at III-D9.
[98] 29 U.S.C. §660(c)(1).
[99] *Id.* at §658.
[100] Fed. R. Civ. P. 4(d)(3).
[101] 29 U.S.C. §659(a).
[102] Schneid, T., Preparing for an OSHA inspection, *Kentucky Manufacturer,* February, 1992.

# chapter two

# Ergonomics in the workplace

> *"All too often we say of a man doing a good job that he is indispensable. A flattering canard, as so many disillusioned and retired and fired have discovered when the world seems to keep on turning without them. In business, a man can come nearest to indispensability by being dispensable in his current job. How can a man move up to new responsibilities if he is the only one able to handle his present tasks? It matters not how small or large the job you now have, if you have trained no one to do it as well, you're not available; you've made your promotion difficult if not impossible."*
>
> —Malcolm Forbes

> *"The employer generally gets the employees he deserves."*
>
> —Sir Walter Bilbey

Ergonomics is the new "buzz" word in the safety profession. Ergonomics is the attempt to address the multitude of cumulative trauma injuries and illnesses, such as carpel tunnel syndrome, which have inflicted substantial pain and disability upon employees performing various industrial and administrative tasks. Although the regulations and methodologies of ergonomics are in flux, prudent safety professionals should become knowledgeable in this important new area and address the potential risks in their workplaces.

Under the general duty clause of 29 C.F.R. 1910, the Occupational Safety and Health Administration (OSHA) has set mandatory guidelines for ergonomics. This is possibly the most complex compliance program, because all facility operations are different and there is no real solution to controlling ergonomic problems. Due to the extreme manual operations in the meat industry, OSHA has hit hard on meat companies to provide an ergonomically safe environment for employees. Therefore, it is essential that the safety professional have a broad mind to understand ergonomics and an imagination to brainstorm possible solutions.

Ergonomics is defined as the science relating the worker, all aspects of the job, and the job environment. The word "ergonomics" breaks down into two words: *Ergo* means the act of work, while *nomics* means the law of work; therefore, ergonomics is the law of work. Several terms are associated with ergonomics, including:

- *Cumulative trauma disorder (CTD):* Disorder that arises from repeated stress and develops as a result of chronic exposure of a particular body part to repeated stress; for example, use of the same, repetitive knifing motion by a meat cutter while boning could result in a CTD.
- *Carpal tunnel syndrome:* Compression of the medial nerve in the carpal tunnel, a passage in the wrist through which finger tendons and a major nerve pass to the hand from the forearm. It is often associated with tingling, pain, or numbness in the thumb and first three fingers.
- *Tendinitis:* Inflammation of the muscle-tendon junction and adjacent muscle tissues resulting from repeated stress of a body member.
- *Tennis elbow:* Inflammation of tissue in the elbow.
- *Trigger finger:* Condition in which the finger frequently flexes against resistance.

# Implementing the program

## Management commitment and employee involvement

The first step in developing an ergonomic program is to gain management commitment. This will provide the organizational resources and motivating force necessary to deal effectively with ergonomic hazards. Employee involvement and feedback are likewise essential for identification of existing and potential hazards and for development and implementation of this program.

## Ergonomic team

The second step in the ergonomic process is to develop an ergonomic team for identifying and correcting ergonomic hazards in the workplace. The team should consist of a wide range of personnel in your facility. Members should include such personnel as members of the safety department, occupational nursing department, production managers, production supervisors, and maintenance personnel.

## Worksite analysis

The third step of the ergonomic process is to perform a worksite analysis. This analysis is performed by the ergonomic team and involves examining and identifying existing hazards or conditions and operations that may create a hazard. During the analysis, it is recommended that medical, safety, and workers' compensation records be evaluated for evidence of cumulative trauma disorders. The worksite analysis should identify those work positions requiring an analysis of ergonomic hazards. This analysis should encompass the following:

- Use of an ergonomic checklist that includes components such as posture, force, repetition, vibration, and various upper extremity factors.
- Identification of work positions that put workers at risk of developing cumulative trauma disorders.
- Verification of low-risk factors for light-duty jobs and restricting awkward work positions.
- Verification of risk factors for work positions which have already been evaluated and corrected to the extent feasible.
- Providing results of worksite analysis for assigning light-duty jobs.
- Re-evaluation of all planned, new, and modified facilities, processes, materials, and equipment to ensure that workplace alterations contribute to reducing or eliminating ergonomic hazards.

It is recommended that periodic surveys be conducted at least annually and/or when operations change and/or the need arises. They should be conducted to identify new deficiencies in work practices or engineering controls and to assess the effects of changes in the work processes.

## *Hazard prevention and control*

Once ergonomic hazards are identified through worksite analysis, the next step is to design measures to prevent and/or control these hazards. Ergonomic hazards are primarily prevented by effective design of the workstation, tools, and the job.

### *Engineering controls*

Engineering techniques are the preferred method of correcting ergonomic hazards. The purpose of engineering controls is to make the job fit the person, not the person fit the job. This can be accomplished by designing or modifying the workstation, work methods, and tools to eliminate excessive exertion and awkwardness.

### *Workstation design*

A workstation should be designed to accommodate the person who actually works on a given job; it is not adequate to design for the average or typical worker. Workstations should be easily adjustable and designed or selected to fit specific tasks so that they are comfortable for the workers using them. The work space should be large enough to allow the full range of required movements, especially where knives, saws, hooks, and similar tools are used.

*Design of work methods*

Work methods should be designed to reduce static, extreme, and awkward postures; repetitive motions; and excessive force. Work method design addresses the content of tasks performed by the workers. It requires analysis of the production system to design or modify tasks to eliminate stressors.

*Tool and handle design*

Tools and handles, if well designed, reduce cumulative trauma disorders. For any tool, a variety of grip sizes should be available to achieve a proper fit and reduce ergonomic risk. The appropriate tool should be used to do a specific job. Tools and handles should be selected to eliminate or minimize the following stressors:

- Chronic muscle contraction or steady force
- Extreme or awkward finger/hand/arm positions
- Repetitive forceful motions
- Tool vibration
- Excessive gripping, pinching, pressing with the hand and fingers

*Work practice controls*

These practices include safe and proper work habits that are understood and followed by managers, supervisors, and workers. The elements include proper work techniques, employee conditioning, regular monitoring, feedback, maintenance, adjustments, modifications, and enforcement.

*Proper work techniques:* These techniques should include appropriate training and practice time for employees, such as:

- Proper cutting techniques, including work methods that improve posture and reduce stress and strain on extremities
- Good knife techniques, including steeling and the regular sharpening of knives
- Correct lifting techniques, including proper body mechanics (e.g., using the legs while lifting, not the back)
- Proper use and maintenance of pneumatic and other power tools
- Correct use of ergonomically designed workstations and fixtures

*New employee conditioning period:* New and returning employees should be gradually integrated into a full work load. Employees should be assigned to an experienced trainer for job training and evaluation during this break-in period.

*Monitoring:* Regular monitoring of the workplace should ensure that employees are continuing to use proper work practices. This monitoring will include a periodic review of techniques and a determination of whether or not procedures in use are those specified; if not, then it should be determined why changes have occurred and whether corrective action is necessary.

*Adjustments and modifications:* These are essential when changes occur in the workplace. Such adjustments could apply to the following:

- Line speeds
- Staffing of positions
- Type, size, weight, or temperature of the product handled

*Administrative controls*

Administrative controls are those that reduce the duration, frequency, and severity of exposures to ergonomic stressors. Those controls include:

- Reducing the total number of repetitions per employee by means of decreasing production rates and limiting overtime work
- Providing rest pauses to relieve fatigued muscle-tendon groups
- Increasing the number of employees assigned to a task, thus alleviating severe conditions, especially while lifting heavy objects
- Job rotation as a preventative measure (the principle of job rotation is to alleviate physical fatigue and stress of a particular set of muscles and tendons by rotating employees among other jobs that use different motions; the ergonomic team should review a job analysis of each operation to ensure that the same muscle–tendon groups are not used)
- Providing sufficient standby and relief personnel for a foreseeable condition on production lines
- Preventative maintenance for mechanical and power tools and equipment, including power saws and knives
- Knife-sharpening program (sharp knives should be readily available)

*Personal protective equipment*

Personal protective equipment (PPE) should be selected with ergonomic stressors in mind. PPE should be provided in a variety of sizes, should accommodate the physical requirements of workers and the job, and should not contribute to ergonomic hazards. The following PPE should be considered for purchase:

- Gloves that facilitate the grasping of tools required for a particular job while protecting the worker from injury
- Apparel that provides protection against extreme temperatures
- Braces, splints, back belts for support

## *Medical management*

Implementation of a medical management system is a major component of an ergonomic program. It is necessary to eliminate and reduce the risk of development of cumulative trauma disorder signs and symptoms through early identification and treatment. The medical management system should address the following:

- Injury and illness recordkeeping
- Early recognition and reporting
- Systematic evaluation and referral
- Conservative treatment
- Conservative return to work
- Systematic monitoring
- Adequate staffing and facilities

*Periodic workplace walkthrough*

If feasible and warranted, members of a medical management team should conduct workplace walkthroughs to remain knowledgeable about operations and work practices to identify potential light-duty jobs and to maintain close contact with employees.

*Symptoms survey*

A survey of employees should be conducted to measure employees' awareness of work-related disorders and to report the location, frequency, and duration of discomfort. Not including employee names on surveys encourages employee participation in the survey. The major strength of the survey is collection of data on the number of workers that may be experiencing cumulative trauma disorders. The survey should be conducted *annually* to help determine any major changes in severity, incidence, and/or location of reported symptoms.

*Symptoms survey checklist*

This checklist should be constructed for your specific industry and operation. It should be distributed to all employees to identify injury and illness. The checklist should aid the ergonomic team in identifying hazard areas.

## *Training and education*

The purpose of training and education is to ensure that employees are sufficiently informed about ergonomic hazards to which they may be exposed and thus are able to participate actively in their own protection. Training and education should provide an overview of potential risks of illness and injuries, their causes and early symptoms, the means of prevention, and treatment. Training allows managers, supervisors, and employees to understand ergonomics and other hazards associated with the job or the production process, their prevention and control, and their medical consequences.

A training program should include the following:

- All affected employees
- Engineers and maintenance personnel

- Supervisors
- Managers
- Medical personnel

*General training*

Employees who are potentially exposed to ergonomic hazards should be given formal instruction on the hazards associated with their jobs and with their equipment. This includes information on the varieties of cumulative trauma disorders, what risk factors cause or contribute to them, how to recognize and report symptoms, and how to prevent these disorders.

*Job-specific training*

New employees and re-assigned workers should receive an initial orientation and hands-on training prior to being placed in a full-time production job. Training production lines may be used for this purpose. Each new hire should receive a demonstration of the proper use of and procedures for all tools and equipment. The initial training program should include the following:

- Care, use, and handling techniques of tools
- Use of special tools and devices associated with individual workstations
- Use of appropriate lifting techniques and devices
- Use of appropriate guards and safety equipment, including personal protective equipment

*Training for supervisors*

Supervisors are responsible for ensuring that employees follow safe work practices. Supervisors should undergo training comparable to that of the employees in addition to further training that will enable them to recognize early signs and symptoms of cumulative trauma disorders, to recognize hazardous work practices, and to correct such practices.

*Training for managers*

Managers should receive training pertaining to ergonomic issues at each workstation and in the production process so they can effectively carry out their responsibilities.

*Training for maintenance*

Maintenance personnel should be trained in prevention and correction of ergonomic hazards through job and workstation design and proper maintenance.

## *Auditing*

The ergonomic team should review the ergonomic program to ensure that new and existing ergonomic hazards are identified and corrected. If the need arises, corrective measures can be taken to eliminate or minimize the hazards.

### *Disciplinary action*

In the event that employees refuse to comply with any safety program, disciplinary action may be required. Employees must understand that top management will not tolerate refusal to abide by this safety program.

## *Ergonomic checklist*

- Has an ergonomic team been developed in the facility?
- Has a worksite analysis been conducted to identify ergonomic hazards?
- Have posture, force, repetition, vibration, and various other factors been observed? If so, has equipment been modified or has PPE been issued to eliminate ergonomic hazards?
- Have periodic surveys been conducted to check for new and existing ergonomic hazards?
- If ergonomic hazards have been identified, have measures to prevent or control these hazards been developed?
- In all operations, does the job fit the person or does the person fit the job?
- Are all work spaces large enough to allow for the full range of required movements?
- Are work methods designed to reduce static, extreme, and awkward postures?
- Has an analysis of the production process been conducted to identify and eliminate ergonomic stressors?
- Are new and returning employees gradually integrated into a full work load?
- Is the workplace monitored to ensure employees continue to use proper work procedures?
- Has overtime work been decreased to limit the total number of repetitions per employee?
- Are employees given breaks to relieve fatigued muscle–tendon groups?
- Has the job rotation principle been implemented to alleviate physical fatigue and stress of a set of muscle–tendon groups?
- Has each job been reviewed by the ergonomic team to identify ergonomic hazards?
- Are sufficient standby personnel available for unforeseeable conditions on production lines?
- Is preventative maintenance performed on tools and equipment to limit the amount of force an employee has to exert to perform a task?
- Is PPE available to employees in a variety of sizes?
- Has a medical management system been implemented to research workers' compensation costs for evidence of CTDs?
- Are employees trained and educated regarding ergonomic hazards that they may encounter?
- Does training provide an overview of potential risks of illnesses and injuries of ergonomic hazards?

- Is specialized training conducted specifically for management, supervisors, and maintenance personnel?
- Is the ergonomic program audited annually?
- Has your ergonomic policy been posted for employee observation?
- Have employees been advised of your facility's disciplinary policy for noncompliance?

# *Proposed standard: ergonomics*

## *General industry*

The Occupational Safety and Health Administration proposes to amend Part 1910 of title 29 of the Code of Federal Regulations as follows:

## *Part 1910 [amended]*

1. New Subpart Y of 29 C.F.R. Part 1910 is added to read as follows:

Subpart Y—Ergonomics Program Standard

Authority: Secs. 6 and 8, Occupational Safety and Health Act, 29 U.S.C. 653, 655, 657, Secretary of Labor's Orders Nos. 12-71 (36 FR 8754), 8-76 (41 FR 25059), 9-83 (48 FR 35736), 1-90 (55 FR 9033), or 6-96 (62 FR 111), as applicable; and 29 C.F.R. Part 1911.

§1910.900

## *Table of contents*

This section is the table of contents for the sections in Subpart Y.

DOES THIS STANDARD APPLY TO ME?

1910.901 Does this standard apply to me?
1910.902 Does this standard allow me to rule out some musculoskeletal disorders (MSDs)?
1910.903 Does this standard apply to the entire workplace or to other workplaces in the company?
1910.904 Are there areas this standard does not cover?

HOW DOES THIS STANDARD APPLY TO ME?

1910.905 What are the elements of a complete ergonomics program?
1910.906 How does this standard apply to manufacturing and manual handling jobs?
1910.907 How does this standard apply to other jobs in general industry?
1910.908 How does this standard apply if I already have an ergonomics program?
1910.909 May I use a Quick Fix instead of setting up a full ergonomics program?
1910.910 What must I do if the Quick Fix does not work?

MANAGEMENT LEADERSHIP AND EMPLOYEE PARTICIPATION

1910.911 What is my basic obligation?
1910.912 What must I do to provide management leadership?
1910.913 What ways must employees have to participate in the ergonomics program?

HAZARD INFORMATION AND REPORTING

1910.914 What is my basic obligation?
1910.915 What information must I provide to employees?
1910.916 What must I do to set up a reporting system?

JOB HAZARD ANALYSIS AND CONTROL

1910.917 What is my basic obligation?
1910.918 What must I do to analyze a problem job?
1910.919 What hazard control steps must I follow?
1910.920 What kinds of controls must I use?
1910.921 How far must I go in eliminating or materially reducing MSD hazards when a covered MSD occurs?
1910.922 What is the "incremental abatement process" for materially reducing MSD hazards?

TRAINING

1910.923 What is my basic obligation?
1910.924 Who must I train?
1910.925 What subjects must training cover?
1910.926 What must I do to ensure that employees understand the training?
1910.927 When must I train employees?
1910.928 Must I retrain employees who have already received training?

MSD MANAGEMENT

1910.929 What is my basic obligation?
1910.930 How must I make MSD management available?
1910.931 What information must I provide to the healthcare professional (HCP)?
1910.932 What must the HCP's written opinion contain?
1910.933 What must I do if temporary work restrictions are needed?
1910.934 How long must I maintain the employee's work restriction protection (WRP) when an employee is on temporary work restrictions?
1910.935 May I offset an employee's WRP if the employee receives workers' compensation or other income?

PROGRAM EVALUATION

1910.936 What is my basic obligation?
1910.937 What must I do to evaluate my ergonomics program?

1910.938 What must I do if the evaluation indicates that my program has deficiencies?

WHAT RECORDS MUST I KEEP?

1910.939 Do I have to keep records of the ergonomics program?
1910.940 What records must I keep and for how long?

WHEN MUST MY PROGRAM BE IN PLACE?

1910.941 When does this standard become effective?
1910.942 When do I have to be in compliance with this standard?
1910.943 What must I do if some or all of the compliance deadlines have passed before a covered MSD is reported?
1910.944 May I discontinue any parts of my ergonomics program if covered MSDs no longer are occurring?

DEFINITIONS

1910.945 What are the key terms in this standard?

*Note to* §1910.900: In this standard, the terms that are defined in §1910.945 are put in "quotations" the first time they appear.

## *DOES THIS STANDARD APPLY TO ME?*

**§1910.901 Does this standard apply to me?**

This standard applies to employers in general industry whose employees work in "manufacturing jobs" or "manual handling jobs," or report "musculoskeletal disorders (MSDs)" that meet the criteria of this standard. This standard applies to the following "jobs:"

(a) *Manufacturing jobs.* Manufacturing jobs are production jobs in which employees perform the "physical work activities" of producing a product and in which these activities make up a significant amount of their worktime;

(b) *Manual handling jobs.* Manual handling jobs are jobs in which employees perform forceful lifting/lowering, pushing/pulling, or carrying. Manual handling jobs include only those jobs in which forceful manual handling is a core element of the employee's job; and

*Note to* §1910.901(a) and (b): Although each manufacturing and manual handling job must be considered on the basis of its actual physical work activities and conditions, the definitions section of this standard (§1910.945) includes a list of jobs that are typically included in and excluded from these definitions.

(c) *Jobs with a musculoskeletal disorder.* Jobs with an MSD are those jobs in which an employee reports an MSD that meets all of these criteria:

(1) The MSD is reported after [the effective date];

(2) The MSD is an "OSHA-recordable MSD," or one that would be recordable if "you" were required to keep OSHA injury and illness records; and

(3) The MSD also meets the screening criteria in §1910.902.

*Note to* §1910.901(c): In this standard, the term "covered MSD" refers to a musculoskeletal disorder that meets the requirements of this section.

**§1910.902 Does this standard allow me to rule out some MSDs?**

Yes. The standard only covers those OSHA-recordable MSDs that also meet these screening criteria:

(a) The physical work activities and conditions in the job are reasonably likely to cause or contribute to the type of MSD reported; and

(b) These activities and conditions are a core element of the job and/or make up a significant amount of the employee's worktime.

**§1910.903 Does this standard apply to the entire workplace or to other workplaces in the company?**

No. This standard is job-based. It only applies to the jobs specified in §1910.901, not to your entire workplace or to other workplaces in your company.

**§1910.904 Are there areas this standard does not cover?**

Yes. This standard does not apply to agriculture, construction, or maritime operations.

## *HOW DOES THIS STANDARD APPLY TO ME?*

**§1910.905 What are the elements of a complete ergonomics program?**

In this standard, a full "ergonomics" program consists of these six program elements:

(a) Management Leadership and Employee Participation;

(b) Hazard Information and Reporting;

(c) Job Hazard Analysis and Control;

(d) Training;

(e) "MSD Management," and

(f) Program Evaluation.

**§1910.906 How does this standard apply to manufacturing and manual handling jobs?**

You must:

(a) Implement the first two elements of the ergonomics program (Management Leadership and Employee Participation, and Hazard Information and Reporting) even if no MSD has occurred in those jobs.

(b) Implement the other program elements when either of the following occurs in those jobs (unless you "eliminate MSD hazards" using the Quick Fix option in §1910.909):

(1) A covered MSD is reported; or

(2) "Persistent MSD symptoms" are reported, plus:

(i) You "have knowledge" that an MSD hazard exists in the job;

(ii) Physical work activities and conditions in the job are reasonably likely to cause or contribute to the type of "MSD symptoms" reported; and

(iii) These activities and conditions are a core element of the job and/or make up a significant amount of the employee's worktime.

*Note to* §1910.906: "Covered MSD" refers to MSDs that meet the criteria in §1910.901(c). As it applies to manufacturing and manual handling jobs, "covered MSD" also refers to persistent MSD symptoms that meet the criteria of this section.

**§1910.907 How does this standard apply to other jobs in general industry?**

In other jobs in general industry, you must comply with all of the program elements in the standard when a covered MSD is reported (unless you eliminate the MSD hazards using the Quick Fix option).

**§1910.908 How does this standard apply if I already have an ergonomics program?**

If you already have an ergonomics program for the jobs this standard covers, you may continue that program, even if it differs from the one this standard requires, provided you show that:

(a) Your program satisfies the basic obligation section of each program element in this standard, and you are in compliance with the recordkeeping requirements of this standard (§§1910.939 and 1910.940);

(b) You have implemented and evaluated your program and controls before [the effective date]; and

(c) The evaluation indicates that the program elements are functioning properly and that you are in compliance with the control requirements in §1910.921.

**§1910.909 May I do a Quick Fix instead of setting up a full ergonomics program?**

Yes. A Quick Fix is a way to fix a "problem job" quickly and completely. If you "eliminate MSD hazards" using a Quick Fix, you do not have to set up the full ergonomics program this standard requires. You must do the following when you Quick Fix a problem job:

(a) Promptly make available the MSD management this standard requires;

(b) Consult with employee(s) in the problem job about the physical work activities or conditions of the job they associate with the difficulties, observe the employee(s) performing the job to identify whether any risk factors are present, and ask employee(s) for recommendations for eliminating the MSD hazard;

(c) Put in Quick Fix controls within 90 days after the covered MSD is identified, and check the job within the next 30 days to determine whether the controls have eliminated the hazard;

(d) Keep a record of the Quick Fix controls; and

(e) Provide the hazard information this standard requires to employee(s) in the problem job within the 90-day period.

*Note to* §1910.909: If you show that the MSD hazards only pose a risk to the employee with the covered MSD, you may limit the Quick Fix to that individual employee's job.

### §1910.910 What must I do if the Quick Fix does not work?

You must set up the complete ergonomics program if either of these occurs:

(a) The Quick Fix controls do not eliminate the MSD hazards within the Quick Fix deadline (120 days); or

(b) Another covered MSD is reported in that job within 36 months.

*Exception:* If a second covered MSD occurs in that job resulting from different physical work activities and conditions, you may use the Quick Fix a second time.

## *MANAGEMENT LEADERSHIP AND EMPLOYEE PARTICIPATION*

### §1910.911 What is my basic obligation?

You must demonstrate management leadership of your ergonomics program. Employees (and their designated representatives) must have ways to report "MSD signs" and "MSD symptoms," get responses to reports, and be involved in developing, implementing, and evaluating each element of your program. You must not have policies or practices that discourage employees from participating in the program or from reporting MSDs signs or symptoms.

### §1910.912 What must I do to provide management leadership?

You must:

(a) Assign and communicate responsibilities for setting up and managing the ergonomics program so managers, supervisors, and employees know what you expect of them and how you will hold them accountable for meeting those responsibilities;

(b) Provide those persons with the authority, "resources," information, and training necessary to meet their responsibilities;

(c) Examine your existing policies and practices to ensure that they encourage and do not discourage reporting and participation in the ergonomics program; and

(d) Communicate "periodically" with employees about the program and their concerns about MSDs.

**§1910.913 What ways must employees have to participate in the ergonomics program?**

Employees (and their designated representatives) must have:

(a) A way to report MSD signs and symptoms;

(b) Prompt responses to their reports;

(c) Access to this standard and to information about the ergonomics program; and

(d) Ways to be involved in developing, implementing, and evaluating each element of the ergonomics program.

## *HAZARD INFORMATION AND REPORTING*

**§1910.914 What is my basic obligation?**

You must set up a way for employees to report MSD signs and symptoms and to get prompt responses. You must evaluate employee reports of MSD signs and symptoms to determine whether a covered MSD has occurred. You must periodically provide information to employees that explains how to identify and report MSD signs and symptoms.

**§1910.915 What information must I provide to employees?**

You must provide this information to current and new employees:

(a) Common MSD hazards;

(b) The signs and symptoms of MSDs and the importance of reporting them early;

(c) How to report MSD signs and symptoms; and

(d) A summary of the requirements of this standard.

**§1910.916 What must I do to set up a reporting system?**

You must:

(a) Identify at least one person to receive and respond to employee reports, and to take the action this standard requires.

(b) Promptly respond to employee reports of MSD signs or symptoms in accordance with this standard.

## JOB HAZARD ANALYSIS AND CONTROL

### §1910.917 What is my basic obligation?

You must analyze the problem job to identify the "ergonomic risk factors" that result in MSD hazards. You must eliminate the MSD hazards, reduce them to the extent feasible, or materially reduce them using the incremental abatement process in this standard. If you show that the MSD hazards only pose a risk to the employee with the covered MSD, you may limit the job hazard analysis and control to that individual employee's job.

### §1910.918 What must I do to analyze a problem job?

You must:

(a) Include in the job hazard analysis all of the employees in the problem job or those who represent the range of physical capabilities of employees in the job;

(b) Ask the employees whether performing the job poses physical difficulties and, if so, which physical work activities or conditions of the job they associate with the difficulties;

(c) Observe the employees performing the job to identify which of the following physical work activities, workplace conditions, and ergonomic risk factors are present:

*Ergonomic risk factors that may be present for physical work activities and conditions*

(1) Exerting considerable physical effort to complete a motion
- (i) Force
- (ii) Awkward postures
- (iii) Contact stress

(2) Doing same motion over and over again
- (i) Repetition
- (ii) Force
- (iii) Awkward postures
- (iv) Cold temperatures

(3) Performing motions constantly without short pauses or breaks in between
- (i) Repetition
- (ii) Force
- (iii) Awkward postures
- (iv) Static postures
- (v) Contact stress
- (vi) Vibration

(4) Performing tasks that involve long reaches
  (i) Awkward postures
  (ii) Static postures
  (iii) Force
(5) Working surfaces that are too high or too low
  (i) Awkward postures
  (ii) Static postures
  (iii) Force
  (iv) Contact stress
(6) Maintaining same position or posture while performing tasks
  (i) Awkward posture
  (ii) Static postures
  (iii) Force
  (iv) Cold temperatures
(7) Sitting for a long time
  (i) Awkward posture
  (ii) Static postures
  (iii) Contact stress
(8) Using hand and power tools
  (i) Force
  (ii) Awkward postures
  (iii) Static postures
  (iv) Contact stress
  (v) Vibration
  (vi) Cold temperatures
(9) Vibrating working surfaces, machinery, or vehicles
  (i) Vibration
  (ii) Force
  (iii) Cold temperatures
(10) Workstation edges or objects that press hard into muscles or tendons
  (i) Contact stress
(11) Using hand as a hammer
  (i) Contact stress
  (ii) Force
(12) Using hands or body as clamp to hold object while performing tasks
  (i) Force
  (ii) Static postures

(iii) Awkward postures
(iv) Contact stress

(13) Gloves that are bulky, too large, or too small
(i) Force
(ii) Contact stress

*Ergonomic risk factors that may be present for manual handling (lifting/lowering, pushing/pulling and carrying)*

(14) Objects or people moved that are heavy
(i) Force
(ii) Repetition
(iii) Awkward postures
(iv) Static postures
(v) Contact stress

(15) Horizontal reach that is long (distance of hands from body to grasp object to be handled)
(i) Force
(ii) Repetition
(iii) Awkward postures
(iv) Static postures
(v) Contact stress

(16) Vertical reach that is below knees or above the shoulders (distance of hands above the ground when the object is grasped or released)
(i) Force
(ii) Repetition
(iii) Awkward postures
(iv) Static postures
(v) Contact stress

(17) Objects or people that are moved significant distance
(i) Force
(ii) Repetition
(iii) Awkward postures
(iv) Static postures
(v) Contact stress

(18) Bending or twisting that occurs during manual handling
(i) Force
(ii) Repetition

(iii) Awkward postures
(iv) Static postures

(19) Object that is slippery or has no handles
(i) Force
(ii) Repetition
(iii) Awkward postures
(iv) Static postures

(20) Floor surfaces that are uneven, slippery or sloped
(i) Force
(ii) Repetition
(iii) Awkward postures
(iv) Static postures

(d) Evaluate the ergonomic risk factors in the job to determine the MSD hazards associated with the covered MSD. As necessary, evaluate the duration, frequency, and magnitude of employee exposure to the risk factors.

**§1910.919 What hazard control steps must I follow?**

You must:

(a) Ask employees in the problem job for recommendations about eliminating or materially reducing the MSD hazards;

(b) Identify, assess, and implement feasible controls (interim and/or permanent) to eliminate or materially reduce the MSD hazards. This includes prioritizing the control of hazards, where necessary;

(c) Track your progress in eliminating or materially reducing the MSD hazards. This includes consulting with employees in problem jobs about whether the implemented controls have eliminated or materially reduced the hazards; and

(d) Identify and evaluate MSD hazards when you change, design, or purchase equipment or change processes in problem jobs.

**§1910.920 What kinds of controls must I use?**

(a) In this standard, you must use any combination of "engineering," "administrative," and/or "work practice controls" to eliminate or materially reduce MSD hazards. Engineering controls, where feasible, are the preferred method for eliminating or materially reducing MSD hazards; however, administrative and work practice controls also may be important in addressing MSD hazards.

(b) "Personal protective equipment" (PPE) may be used to supplement engineering, work practice, and administrative controls, but may only be used alone where other controls are not feasible. Where PPE is used, you must provide it at "no cost to employees."

*Note to* §1910.920: Back belts/braces and wrist braces/splints are not considered PPE for the purposes of this standard.

**§1910.921 How far must I go in eliminating or materially reducing MSD hazards when a covered MSD occurs?**

The occurrence of a covered MSD in a problem job is not itself a violation of this standard. You must comply with one of the following:

(a) You implement controls that materially reduce the MSD hazards using the incremental abatement process in §1910.922; or

*Note to* §1910.921(a): "Materially reduce MSD hazards" means to reduce the duration, frequency, and/or magnitude of exposure to one or more ergonomic risk factors in a way that is reasonably anticipated to significantly reduce the likelihood that covered MSDs will occur.

(b) You implement controls that reduce the MSD hazards to the extent feasible. Then, you periodically look to see whether additional controls are now feasible and, if so, you implement them promptly; or

(c) You implement controls that eliminate the MSD hazards in the problem job.

*Note to* §1910.921(c): "Eliminate MSD hazards" means that you eliminate employee exposure to ergonomic risk factors associated with the covered MSD or you reduce employee exposure to the risk factors to such a degree that a covered MSD is no longer reasonably likely to occur.

**§1910.922 What is the "incremental abatement process" for materially reducing MSD hazards?**

You may materially reduce MSD hazards using the following incremental abatement process:

(a) When a covered MSD occurs, you implement one or more controls that materially reduce the MSD hazards; and

(b) If continued exposure to MSD hazards in the job prevents the injured employee's condition from improving or another covered MSD occurs in that job, you implement additional feasible controls to materially reduce the hazard further; and

(c) You do not have to put in further controls if the injured employee's condition improves and no additional covered MSD occurs in the job. However, if the employee's condition does not improve or another covered MSD occurs, you must continue this incremental abatement process if other feasible controls are available.

## *TRAINING*

**§1910.923 What is my basic obligation?**

You must provide training to employees so they know about MSD hazards and your ergonomics program and measures for eliminating or materially reducing the hazards. You must provide training initially, periodically, and at least every 3 years at no cost to employees.

**§1910.924 Who must I train?**

You must train:

(a) Employees in problem jobs;

(b) Supervisors of employees in problem jobs; and

(c) Persons involved in setting up and managing the ergonomics program, except for any outside consultant you may use.

**§1910.925 What subjects must training cover?**

This table specifies the subjects training must cover:

| *You must provide training for...* | *so that they know...* |
|---|---|
| (a) Employees in problem jobs and their supervisors | How to recognize MSD signs and symptoms; |
| | How to report MSD signs and symptoms and the importance of early reporting; |
| | MSD hazards in their jobs and the measures they must follow to protect themselves from exposure to MSD hazards; |
| | Job-specific controls implemented in their jobs; |
| | The ergonomics program and their role in it; and |
| | The requirements of this standard. |
| (b) Persons involved in setting up and managing the ergonomics program | The subjects above; |
| | How to set up and manage an ergonomics program; |
| | How to identify and analyze MSD hazards and measures to eliminate or materially reduce the hazards; and |
| | How to evaluate the effectiveness of ergonomics programs and controls. |

**§1910.926 What must I do to ensure that employees understand the training?**

You must provide training and information in language that employees understand. You also must give employees an opportunity to ask questions and receive answers.

**§1910.927 When must I train employees?**

The following table specifies when you must train employees:

| *If you have...* | *then you must provide training at these times...* |
|---|---|
| (a) Employees in problem jobs and their supervisors | When a problem job is identified;<br>When initially assigned to a problem job;<br>Periodically as needed (e.g., when new hazards are identified in a problem job or changes are made to a problem job that may increase exposure to MSD hazards); and<br>At least every 3 years. |
| (b) Persons involved in setting up and managing the ergonomics program | When they are initially assigned to setting up and managing the ergonomics program;<br>Periodically as needed (e.g., when evaluation reveals significant deficiencies in the program, when significant changes are made in the ergonomics program); and<br>At least every 3 years. |

### §1910.928 Must I retrain employees who have received training already?

No. You do not have to provide initial training to current employees, new employees, and persons involved in setting up and managing the ergonomics programs if they have received training in the subjects this standard requires within the last 3 years. However, you must provide initial training in the subjects in which they have not been trained.

## *MSD MANAGEMENT*

### §1910.929 What is my basic obligation?

You must make MSD management available promptly whenever a covered MSD occurs. You must provide MSD management at no cost to employees. You must provide employees with the temporary "work restrictions" and "work restriction protection" (WRP) this standard requires.

### §1910.930 How must I make MSD management available?

You must:

(a) Respond promptly to employees with covered MSDs to prevent their condition from getting worse;

(b) Promptly determine whether temporary work restrictions or other measures are necessary;

(c) When necessary, provide employees with prompt access to a "healthcare professional" (HCP) for evaluation, management, and "follow-up;"

(d) Provide the HCP with the information necessary for conducting MSD management; and

(e) Obtain a written opinion from the HCP and ensure that the employee is also promptly provided with it.

**§1910.931 What information must I provide to the healthcare professional?**

You must provide:

(a) A description of the employee's job and information about the MSD hazards in it;

(b) A description of available work restrictions that are reasonably likely to fit the employee's capabilities during the recovery period;

(c) A copy of this MSD management section and a summary of the requirements of this standard;

(d) Opportunities to conduct workplace walk-throughs.

**§1910.932 What must the HCP's written opinion contain?**

The written opinion must contain:

(a) The HCP's opinion about the employee's medical conditions related to the MSD hazard in the employee's job.

(1) You must instruct the HCP that any findings, diagnoses, or information not related to workplace exposure to MSD hazards must remain confidential and must not be put in the written opinion or communicated to you.

(2) To the extent permitted and required by law, you must ensure employee privacy and confidentiality regarding medical conditions related to workplace exposure to MSD hazards that are identified during the MSD management process;

(b) Any recommended temporary work restrictions and follow-up;

(c) A statement that the HCP informed the employee about the results of the evaluation and any medical conditions resulting from exposure to MSD hazards that require further evaluation or treatment;

(d) A statement that the HCP informed the employee about other physical activities that could aggravate the covered MSD during the recovery period.

**§1910.933 What must I do if temporary work restrictions are needed?**

You must:

(a) Work restrictions: Provide temporary work restrictions, where necessary, to employees with covered MSDs. Where you have referred the employee to a HCP, you must follow the temporary work restriction recommendations in the HCP's written opinion;

(b) Follow-up: Ensure that appropriate follow-up is provided during the recovery period; and

(c) Work restriction protection (WRP): Maintain the employee's WRP while temporary work restrictions are provided. You may condition the provision of WRP on the employee's participation in the MSD management this standard requires.

**§1910.934 How long must I maintain the employee's work restriction protection when an employee is on temporary work restriction?**

You must maintain the employee's WRP until the *first* of these occurs:

(a) The employee is determined to be able to return to the job;

(b) You implement measures that eliminate the MSD hazards or materially reduce them to the extent that the job does not pose a risk of harm to the injured employee during the recovery period; or

(c) 6 months have passed.

**§1910.935 May I offset an employee's WRP if the employee receives workers' compensation or other income?**

Yes. You may reduce the employee's WRP by the amount the employee receives during the work restriction period from:

(a) Workers' compensation payments for lost earnings;

(b) Payments for lost earnings from a compensation or insurance program that is publicly funded or funded by you; and

(c) Income from a job taken with another employer that was made possible because of the work restrictions.

## *PROGRAM EVALUATION*

**§1910.936 What is my basic obligation?**

You must evaluate your ergonomics program periodically, and at least every 3 years, to ensure that it is in compliance with this standard.

**§1910.937 What must I do to evaluate my ergonomics program?**

You must:

(a) Consult with employees in problem jobs to assess their views on the effectiveness of the program and to identify any significant deficiencies in the program;

(b) Evaluate the elements of your program to ensure they are functioning properly; and

(c) Evaluate the program to ensure it is eliminating or materially reducing MSD hazards.

**§1910.938 What must I do if the evaluation indicates my program has deficiencies?**

If your evaluation indicates that your program has deficiencies, you must promptly take action to correct those deficiencies so that your program is in compliance with this standard.

## WHAT RECORDS MUST I KEEP?

**§1910.939 Do I have to keep records of the ergonomics program?**

You only have to keep records if you had 10 or more employees (including part-time employees and employees provided through personnel services) on any one day during the preceding calendar year.

**§1910.940 What records must I keep and for how long?**

This table specifies the records you must keep and how long you must keep them:

| *You must keep these records...* | *for at least...* |
|---|---|
| (a) Employee reports and your responses | 3 years. |
| (b) Job hazard analysis | 3 years. |
| (c) Hazard control records | 3 years. |
| (d) Quick Fix control records | 3 years. |
| (e) Ergonomics program evaluation | 3 years, or until replaced by updated records, whichever comes first. |
| (f) MSD management records | The duration of the injured employee's employment plus 3 years. |

*Note to* §1910.940: The record retention period in this standard is shorter than that required by OSHA's rule on Access to Employee Exposure and Medical Records (29 C.F.R. 1910.1020). However, you must comply with the other requirements of that rule.

## WHEN MUST MY PROGRAM BE IN PLACE?

**§1910.941 When does this standard become effective?**

This standard becomes effective 60 days after [publication date of final rule].

**§1910.942 When do I have to be in compliance with this standard?**

This standard provides start-up time for setting up the ergonomics program and putting in controls in problem jobs. You must comply with the requirements of this standard, including recordkeeping, by the deadlines in the table on the following page.

*Note to* §1910.942: The compliance deadlines in this section do not apply if you are using a Quick Fix.

| *You must comply with these requirements and related recordkeeping...* | *no later than... .* |
|---|---|
| (a) MSD management | Promptly when an MSD is reported. |
| (b) Management leadership and employee participation | 1 year after the effective date. |
| (c) Hazard information and reporting | 1 year after the effective date. |
| (d) Job hazard analysis | 1 year after the effective date. |
| (e) Interim controls | 1 year after the effective date. |
| (f) Training | 2 years after the effective date. |
| (g) Permanent controls | 2 years after the effective date. |
| (h) Program evaluation | 3 years after the effective date. |

**§1910.943 What must I do if some or all of the compliance deadlines have passed before a covered MSD is reported?**

If the compliance start-up deadline has passed before you must comply with a particular element of this standard, you may take the following additional time to comply with that element and the related recordkeeping:

| *You must comply with these requirements and related recordkeeping...* | *within...* |
|---|---|
| (a) MSD management | 5 days. |
| (b) Management leadership and employee participation | 5 days. |
| (c) Hazard information and reporting | 30 days (in manufacturing and manual handling jobs, these requirements must be implemented by [1 year after the effective date]). |
| (d) Job hazard analysis | 60 days. |
| (e) Interim controls | 60 days. |
| (f) Training | 90 days. |
| (g) Permanent controls | 90 days. |
| (h) Program evaluation | 1 year. |

*Note to* §1910.943: The compliance deadlines in this section do not apply if you are using a Quick Fix.

**§1910.944 May I discontinue certain aspects of my program if covered MSDs no longer are occurring?**

Yes. However, as long as covered MSDs are reported in a job, you must maintain all the elements of the ergonomics program for that job. If you eliminate or materially reduce the MSD hazards and no covered MSD is reported for 3 years, you only have to continue the elements in this table:

| *If you eliminate or materially reduce the hazards and no covered MSD is reported for 3 years in...* | *then you may stop all except the following parts of your program in that job...* |
|---|---|
| (a) A manufacturing or manual handling job | Management leadership and employee participation<br>Hazard information and reporting<br>Maintenance of implemented controls and training related to the controls |
| (b) Other jobs in general industry where a covered MSD had been reported | Maintenance of controls and training related to the controls |

## DEFINITIONS

### §1910.945 What are the key terms in this standard?

"**Administrative controls**" are changes in the way that work in a job is assigned or scheduled that reduce the magnitude, frequency, or duration of exposure to ergonomic risk factors. Examples of administrative controls for MSD hazards include:

- Employee rotation
- Job task enlargement
- Alternative tasks
- Employer-authorized changes in work place

A "**covered MSD**" is an MSD, reported in any job in general industry, that meets these criteria:

- It is reported after [the effective date].
- It is an OSHA-recordable MSD.
- It occurred in a job in which the physical work activities and conditions are reasonably likely to cause or contribute to the type of MSD reported.
- These activities and conditions are a core element and/or make up a significant amount of the employee's worktime.

In a manufacturing or manual handling job, persistent MSD symptoms are also considered a covered MSD if they meet these criteria:

- They last for at least 7 consecutive days after they are reported.
- The employer has knowledge that an MSD hazard exists in the job.

- They occurred in a job in which the physical work activities and conditions are reasonably likely to cause or contribute to the type of MSD signs or symptoms reported.
- These activities and conditions are a core element and/or make up a significant amount of the employee's worktime.

"**Eliminate MSD hazards**" means to eliminate employee exposure to the ergonomic risk factors associated with the covered MSD, or to reduce employee exposure to the risk factors to such a degree that a covered MSD is no longer reasonably likely to occur.

"**Engineering controls**" are physical changes to a job that eliminate or materially reduce the presence of MSD hazards. Examples of engineering controls for MSD hazards include changing, modifying, or redesigning the following:

- Workstations
- Tools
- Facilities
- Equipment
- Materials
- Processes

"**Ergonomics**" is the science of fitting jobs to people. Ergonomics encompasses the body of knowledge about physical abilities and limitations, as well as other human characteristics that are relevant to job design. Ergonomic design is the application of this body of knowledge to the design of the workplace (i.e., work tasks, equipment, environment) for safe and efficient use by workers.

"**Ergonomic risk factors**" are the following aspects of a job that pose a biomechanical stress to the worker:

- Force (i.e., forceful exertions, including dynamic motions)
- Repetition
- Awkward postures
- Static postures
- Contact stress
- Vibration
- Cold temperatures

Ergonomic risk factors are elements of MSD hazards that must be considered in light of their combined effect in causing or contributing to an MSD. Jobs that have multiple risk factors have a greater likelihood of causing or contributing to MSDs, depending on the duration, frequency, and magnitude of employee exposure to each risk factor or to a combination of them. Ergonomic risk factors are also called "ergonomic stressors" and "ergonomic factors."

"**Follow-up**" is the process or protocol an employer and/or HCP uses to check up on the condition of employees with covered MSDs when they are given temporary work restrictions during the recovery period. Prompt follow-up helps to ensure that the MSD is resolving and, if it is not, that other measures are promptly taken.

"**Have knowledge**" means that you have been provided information that MSD hazards exist in a manufacturing or manual handling job by any of the following:

- Insurance company
- Consultant
- Healthcare professional
- Person or persons working for you who have the requisite training to identify and analyze MSD hazards

"**Healthcare professionals**" (HCPs) are physicians or other licensed healthcare professionals whose legally permitted scope of practice (e.g., license, registration, or certification) allows them to independently provide or be delegated the responsibility to provide some or all of the MSD management requirements of this standard.

"**Job**" means the physical work activities or tasks that employees perform. In this standard, the term "job" also includes those jobs involving the same physical work activities and conditions even if the jobs have different titles or classification.

"**Manual handling jobs**" are jobs in which employees perform forceful lifting/lowering, pushing/pulling, or carrying. Manual handling jobs include only those jobs in which forceful manual handling is a core element of an employee's job. Although each job must be considered on the basis of its actual physical work conditions and work activities, this table lists jobs that typically are included in and excluded from this definition:

1. Examples of jobs that typically are manual handling jobs
   - Patient handling jobs (e.g., nurses' aides, orderlies, nurse assistants)
   - Package sorting, handling, and delivering
   - Hand packing and packaging
   - Baggage handling (e.g., porters, airline baggage handlers, airline check-in)
   - Warehouse manual picking and placing
   - Beverage delivering and handling
   - Stock handling and bagging
   - Grocery store bagging
   - Grocery store stocking
   - Garbage collecting
2. Examples of job/tasks that typically are not manual handling jobs
   - Administrative jobs
   - Clerical jobs

- Supervisory/managerial jobs that do not involve manual handling work
- Technical and professional jobs
- Jobs involving unexpected manual handling
- Lifting object or person in emergency situation (e.g., lifting or carrying injured coworker)
- Jobs involving manual handling that is so infrequent it does not occur on any predictable basis (e.g., filling in on a job due to unexpected circumstances, replacing empty water bottle, lifting of box of copier paper)
- Jobs involving manual handling that is done only on an infrequent "as needed" basis (e.g., assisting with delivery of large or heavy package, filling in once for an absent employee)
- Jobs involving minor manual handling that is incidental to the job (e.g., carrying briefcase to meeting or baggage on work travel)

"**Manufacturing jobs**" are production jobs in which employees perform the physical work activities of producing a product and in which these activities make up a significant amount of their worktime. Although each job must be considered on the basis of its actual physical work conditions and work activities, this table lists jobs that typically are included in and excluded from this definition:

1. Examples of jobs that typically are manufacturing jobs
   - Assembly-line jobs producing products (durable and nondurable), subassemblies, components and parts
   - Paced assembly jobs (assembling and disassembling)
   - Piecework assembly jobs (assembling and disassembling) and other time-critical assembly jobs
   - Product inspection jobs (e.g., testers, weighers)
   - Meat, poultry, and fish cutting and packing
   - Machine operation
   - Machine loading/unloading
   - Apparel manufacturing jobs
   - Food preparation assembly-line jobs
   - Commercial baking jobs
   - Cabinetmaking
   - Tire building
2. Examples of jobs that typically are not manufacturing jobs
   - Administrative jobs
   - Clerical jobs
   - Supervisory/managerial jobs that do not involve production work
   - Warehouse jobs in manufacturing facilities
   - Technical and professional jobs
   - Analysts and programmers
   - Sales and marketing

- Procurement/purchasing jobs
- Customer service jobs
- Mailroom jobs
- Security guards
- Cafeteria jobs
- Groundskeeping jobs (e.g., gardeners)
- Jobs in power plant in manufacturing facility
- Janitorial
- Maintenance
- Logging jobs
- Production of food products (e.g., bakery, candy, and other confectionery products) primarily for direct sale on the premises to household customers

"**Materially reduce MSD hazards**" means to reduce the duration, frequency, and/or magnitude of exposure to one or more ergonomic risk factors in a way that is reasonably anticipated to significantly reduce the likelihood that covered MSDs will occur.

"**Musculoskeletal disorders**" (MSDs) are injuries and disorders of the muscles, nerves, tendons, ligaments, joints, cartilage, and spinal discs. Exposure to physical work activities and conditions that involve risk factors may cause or contribute to MSDs. MSDs do not include injuries caused by slips, trips, falls, or other similar accidents. Examples of MSDs include:

- Carpal tunnel syndrome
- Rotator cuff syndrome
- de Quervain's disease
- Trigger finger
- Tarsal tunnel syndrome
- Sciatica
- Epicondylitis
- Tendinitis
- Raynaud's phenomenon
- Carpet-layer's knee
- Herniated spinal disc
- Low back pain

"**MSD hazards**" are physical work activities and/or physical work conditions, in which ergonomic risk factors are present, that are reasonably likely to cause or contribute to a covered MSD.

"**MSD management**" is your process for ensuring that employees with covered MSDs receive prompt and effective evaluation, management, and follow-up, at no cost to them, in order to prevent permanent damage or disability from occurring. In this standard, the MSD management process includes:

1. Evaluation, management, and follow-up of injured employees by persons in the workplace and/or by HCPs
2. A method for identifying available work restrictions and promptly providing them when needed

MSD management does not include establishing specific medical treatment for MSDs. Medical treatment protocols and procedures are established by the healthcare professions.

"**MSD signs**" are objective physical findings that an employee may be developing an MSD. Examples of MSD signs include:

- Decreased range of motion
- Deformity
- Decreased grip strength
- Loss of function

"**MSD symptoms**" are physical indications that an employee may be developing an MSD. Symptoms can vary in severity, depending on the amount of exposure to MSD hazards. Symptoms often appear gradually as muscle fatigue or pain at work that disappears during rest. Symptoms usually become more severe as exposure continues (e.g., tingling continues after work ends, numbness makes it difficult to perform the job, and finally pain is so severe the employee cannot perform the job). MSD symptoms include:

- Numbness
- Burning
- Pain
- Tingling
- Cramping
- Stiffness

"**No cost to employees**" means that PPE, training, MSD management, and other requirements of this standard are provided to employees free of charge and while they are "on the clock" (e.g., paying for time employees spend receiving training outside the work day).

"**OSHA-recordable MSD**" is an MSD that meets the occupational injury and illness recording requirements of 29 C.F.R. Part 1904. Under Part 1904, an MSD is recordable when:

1. Exposure at work caused or contributed to the MSD or aggravated a pre-existing MSD.
2. The MSD results in at least one of the following:
   a. A diagnosis of an MSD by an HCP
   b. A positive physical finding (e.g., an MSD sign or a positive Finkelstein's, Phalen's, or Tinel's test result)

c. An MSD symptom, plus at least one of these:
   - Medical treatment
   - One or more lost work days
   - Restricted work activity
   - Transfer or rotation to another job

"**Periodically**" means that a process or activity, such as records review or training, is performed on a regular basis that is appropriate for the conditions in the workplace. Periodically also means that the process or activity is conducted as often as needed, such as when significant changes are made in the workplace that may result in increased exposure to MSD hazards.

"**Persistent MSD symptoms**" are MSD symptoms that persist for at least 7 consecutive days after they are reported.

"**Personal protective equipment**" (PPE) is equipment employees wear that provides an effective protective barrier between the employee and MSD hazards. Examples of PPE are vibration-reduction gloves and carpet-layer's knee pads.

"**Physical work activities**" are the physical demands, exertions, and functions of the task or job.

"**Problem job**" is a job in which a covered MSD is reported. A problem job also includes any job in the workplace that involves the same physical work activities and conditions as the one in which the covered MSD is reported, even if the jobs have different titles or classifications.

"**Resources**" are the provisions necessary to develop, implement, and maintain an effective ergonomics program. Resources include money (e.g., to purchase items such as job hazard analysis equipment, training materials, and controls), personnel, and worktime to conduct program responsibilities (e.g., job hazard analysis, program evaluation).

"**Work practice controls**" are changes in the way an employee performs the physical work activities of a job that reduce exposure to MSD hazards. Work practice controls involve procedures and methods for safe work. Examples of work practice controls for MSD hazards include:

- Training in proper work postures
- Training in use of the appropriate tool
- Employer-authorized micro-breaks

"**Work restriction protection**" (WRP) means the maintenance of earnings and other employment rights and benefits of employees who are on temporary work restrictions as though they had not been placed on temporary work restriction. For employees who are on restricted work activity, WRP includes maintaining 100% of the after-tax earnings that employees with covered MSDs were receiving at the time they were placed on restricted work activity. For employees who have been removed from the workplace,

WRP includes maintaining 90% of the after-tax earnings. Benefits mean 100% of the non-wage-and-salary value employees were receiving at the time they were placed on restricted work activity or were removed from the workplace. Benefits include seniority, insurance programs, retirement benefits, and savings plans.

"**Work restrictions**" are limitations on an injured employee's exposure to MSD hazards during the recovery period. Work restrictions may involve limitations on the work activities of the employee's current job, transfer to temporary alternative-duty jobs, or complete removal from the workplace. To be effective, work restrictions must not expose the injured employee to the same MSD hazards as were present in the job giving rise to the covered MSD.

"**You**" means the employer as defined by the Occupational Safety and Health Act of 1970 (29 U.S.C. 651 et seq.).

In summation, cumulative trauma disorders cost employers substantial dollars, and the injuries and illnesses incurred by employees are costing the employee dollars as well as causing pain, loss of life activities, and other harms. Safety professionals should address the potential ergonomic risks in their workplace and take appropriate measures to eliminate or minimize these risks. It is highly recommended that safety professionals do not wait for the final promulgation of the OSHA standard or the outcome of any litigation to address the ergonomic needs of their workplace. Development of an effective ergonomics program has become a "must" for many employers, and the dividends that can be derived from an effective ergonomics program for employees and the company can be substantial. Please do not wait ... get started with your ergonomics program!

# *chapter three*

# *Respiratory hazards*

*"Every man must scratch his head with his own nails."*
—Arabian proverb

*"No bird soars too high, if he soars on his own wings."*
—William Blake

Failure to respect respiratory hazards can kill workers in a matter of seconds. Poor work practices can also allow toxic materials to slowly accumulate, and their effects may surface many years later in the form of chronic lung illness, cancers, or diseases of specific target organs. It is imperative that all employees fully understand the physiological aspects of the respiratory system, limitations of the various personal protective equipment (PPE) available, and responses of the body to respiratory hazards. If the employees understand their bodies, the limitations of the personal protective equipment, and their personal limitations, they can make intelligent and safe decisions to protect themselves.

Improperly compensated risk is a primary factor in many unnecessary respiratory hazard incidents. The employees do not have enough information about the material they are being exposed to or they may not understand the limitations of their PPE. Although experience can be a good teacher, it can also reinforce negative behaviors. An employee who does not wear a respirator and completes the task with no apparent ill effects from the exposure to toxic gases, vapors, or airborne particulate hazards may be more likely to repeat the dangerous practice.

At a firefighter safety and health symposium, a cancer researcher described common carcinogens firefighters are exposed to in the line of duty. After learning about the dangers of asbestos, a group of firefighters described a daily practice in their station of removing an asbestos fire blanket from an apparatus compartment to gain access to an engine-powered rotary saw. After removing the blanket and rescue saw, they would place the saw on the ground near the blanket and start the saw to ensure that it would run if needed on their tour of duty. This procedure was performed every morning and again by the evening shift when they relieved the day shift. The researcher shuddered as they talked. Each time the blanket was handled, friable pieces of the asbestos blanket were undoubtedly suspended in the air in the fire house. When the saw was started in close proximity to the blanket, additional fibers were blown all over the station by the engine exhaust. The firefighters did not have a death wish. They just failed to realize the risk they were taking. It is the employer's responsibility under OSHA Act 29 C.F.R. 1910.134 to make sure everyone in its facilities understands all the respiratory hazards and follows safe work practices.

All employees may not be physically fit to wear certain respirators. It is the employer's responsibility to provide medical screening for all employees who may have to wear respirators. OSHA has developed a simple questionnaire that can be administered to determine if a particular employee must be screened by a physician. A copy of that questionnaire is included in Appendix C.

## *Respiratory hazard exposure prevention methods*

Administrative controls are steps that management takes to remove or isolate workers from exposure to certain hazards. Policies, procedures, signs, and permit systems are examples of administrative control methods used to prevent employees from entering certain areas where hazards exist. These same methods may be employed to allow entry into certain areas only after the area(s) have been determined to be safe; also, other means to protect the worker have been approved and documented. A confined-space entry permit system is an example of an administrative control designed to prevent worker exposure and injury.

Engineering controls are those steps taken that attempt to protect workers by isolating them from hazards by using barricades, machine guards, light curtains, or other devices that prevent or minimize exposure to the hazard. A common engineering control used for controlling respiratory hazards is a ventilation hood system. A fan system pulls toxic vapors away from the employee workstation and exhausts them in a safe area.

In controlled industrial hazard settings, personal protective equipment (PPE) is usually the last line of defense. If the hazard cannot be isolated using administrative controls and/or engineering controls, then personal protective equipment is selected specifically for the type of hazard. If the concentration of the hazard present in the air can be measured, an appropriate respirator with a suitable protection factor is selected to ensure that employees are not exposed above acceptable levels. If the exact hazard is unknown, if multiple hazards exist, or if the substance is highly toxic and air-purifying respirators will not provide adequate safety, air-supplied respirators are required. Firefighters, hazardous materials teams, and other first responders rarely have the ability to measure the quantity of substance(s) in the air, so their first choice is usually the positive-pressure, self-contained breathing apparatus (SCBA). These full-facepiece respirators have high-pressure cylinders that supply breathing gas independent of the hazardous atmosphere. They offer the highest level of protection available. Quantitative fit testing should be performed to ensure that these individuals are able to wear the SCBA without leaks around the sealing edge of the facepiece. Most respirator manufacturers offer facepieces in several sizes to accommodate a wide variety of face shapes.

## *Classification of respiratory hazards: gases and fumes*

Simple asphyxiants are gases that displace the air that contains the oxygen required for life. All of the inert gases, such as carbon dioxide, argon, nitrogen, and helium, are examples of simple asphyxiants. Gases with a vapor density heavier than air (greater than 1.0) pose potential hazards to workers in basements and other areas such as confined spaces. Gases with a vapor density lighter than air (less than 1.0) can accumulate at the top of shafts or ceilings. One of the major problems associated with these simple asphyxiants is the fact that they may be odorless, colorless, and tasteless. A worker can enter a space filled with these gases and collapse after inhaling a single breath with no oxygen.

Toxic gases are those that cause problems in very low concentrations. Carbon monoxide is considered immediately dangerous to life and health at concentrations of 200 parts per million. If the particular gas or vapor is also flammable, you need to determine the upper and lower flammable limits. Methane, for example, has a lower flammable limit of 5% and an upper flammable limit of 15%. This means that outside of that narrow 10% range, the mixture will either be too lean (below 5%) or too rich (above 15%) to burn. Some people erroneously believe it is safe to work in concentrations too rich to ignite. Those people should ask themselves two very important questions:

1. How do you move from an area outside the hazard that is too lean to the hazard area that is too rich without going through an area where ignition is possible?
2. If you are working in an area that is too rich for ignition and the atmosphere changes quickly and dilutes the mixture into the flammable range, are you prepared for an extended stay in a burn unit or worse?

The bottom line is do not enter areas containing more than 10–20% of the lower flammable limit of the gas being measured. Also, make sure you either calibrate your combustible gas instrument with the gas being tested or use a chart furnished with the instrument to interpret the results on the challenge gas based upon your calibration gas. When selecting combustible gas instruments, choose continuous or intermittent monitors over spot analyzers, which only provide color change or visual meter indications. Continuous or intermittent monitoring units sample and interpret the atmosphere when workers in the area are busy and will sound an alarm and flash lights to warn the incident commander, thus allowing personnel time to egress from the space before ignition.

Particulates are particles of solid material suspended in the air. These can be inhaled and trapped in the respiratory system. The body has several lines of defense against these, but their ability is limited and should not be relied upon to protect the worker. Almost any type of dust can be carried deep into the lungs and cause irreversible damage to delicate airways and alveoli. Chronic obstructed pulmonary disease (COPD) is treatable if diagnosed early and the exposure is stopped. "Black lung," which is a form of COPD, has ruined the lives of many coalminers exposed to the coal dust. Other workers in many industrial settings are exposed to other dusts that will yield the same result. In advanced stages of COPD, the patient's lungs are unable to exchange oxygen and carbon dioxide. Eventually, the patient's lungs begin to fill with fluid; some eyewitness accounts describe the patient's death as one of drowning in fluid trapped in the patient's lungs. Smoke from tobacco products such as cigarettes causes the lungs to lose their elasticity and compounds the problems of this disease. Employer dollars spent on stopping smoking will come back to the employer in increased worker productivity, less absenteeism, savings on healthcare costs, and improved quality of life for workers. Several fire departments in the U.S. require new hires to sign agreements indicating they will not use tobacco products.

Another common industrial particulate is silica. More than 1 million U.S. workers are exposed to crystalline silica. Each year, more than 250 American workers die from silicosis. There is no cure for the disease, but it is 100% preventable if employers, workers, and health professionals work together to reduce exposures. The smaller the silica particle size, the deeper into the human airway it travels. The OSHA standard specifically addresses this by limiting the total allowable silica dust and respirable silica dust.

Asbestosis is another deadly airborne particulate. An estimated 1.3 million employees in construction and general industry face significant asbestos exposure on the job. Heaviest exposures occur in the construction industry, particularly during removal of asbestos during renovation or demolition. Asbestos exposure is highly regulated by both OSHA and the Environmental Protection Agency (EPA). Asbestos fibers can have serious effects on a person's health if inhaled. There is no known safe exposure to asbestos. The greater the exposure, the greater the risk of developing an asbestos-related disease. The amount of time between initial exposure to asbestos and the first signs of disease can be as long as 30 years. Smokers exposed to asbestos have a higher risk of developing lung cancer than from just smoking alone. Workers who are exposed to asbestos products are often affected with asbestosis, a scarring of the lungs that leads to breathing problems and heart failure. Inhalation of asbestos fibers can cause mesothelioma, a rare cancer of the linings of the chest and abdomen. The American Lung Association indicates it may also be linked to cancers of the stomach, intestines, and rectum. Again, the good news is that these diseases can be prevented through following safe work practices and proper use of PPE.

Environmental considerations can impact the performance of the respirator and the person using it. In extremely cold conditions, the edges of the facepiece may become rigid and fail to maintain an adequate seal, allowing toxic gases to be inhaled. If the breathing air in the cylinders is not processed to Compressed Gas Association Breathing Air Standards of Grade D or better, any moisture in the air can freeze into ice bullets and cause regulators to fail. In extremely toxic environments this can be fatal. In high-temperature environments, the wearer's perspiration may allow the facepiece to slide and loosen the seal between the face and mask. The breathing gas in the SCBA cylinder remains the firefighter's only means for surviving in high-temperature environments where a single breath of superheated air would blister his fragile lungs. The hazardous materials technicians and specialists also work in environments where a single breath of a toxic atmosphere could kill them. This lifesaving air supply is not delivered to the wearer without exacting a price. SCBA rated with a 30-minute service life weighs more than 20 pounds, adding to the physical stress on the body. In situations requiring 60 minutes' or greater duration, the SCBA can weigh as much as 34 pounds. Whether a Level A encapsulating chemical suit or a firefighter's standard turnout clothing is worn, the wearer must be in excellent physical condition to work safely and effectively.

## *Different types of respirators available*

The Occupational Safety and Health Administration has elected to recognize and approve only respirators approved by the National Institute for Occupational Safety and Health (NIOSH) or the Mine Safety and Health Administration (MSHA) as set forth in 42 C.F.R. Part 84.

Air-purifying respirators (APRs) are available in a wide range of styles but all of them attempt to filter toxins from the environment or absorb chemicals as the air passes through the filter medium. APRs are grouped into one of three classes: Particulate-removing APRs utilize filters to reduce inhaled concentrations of dusts, mists, fumes, and/or fibers. Vapor- and gas-removing APRs are designed with sorbent cartridges or canisters that absorb the vapors or fumes before they reach the wearer's airway. The third style of APR combines particulate and vapor/gas removal in a single unit.

Air-purifying respirators can be either nonpowered or powered air purifying, with blowers to pull the contaminated air through the filter/canister. APRs can be single use (disposable) or can utilize replaceable filters and/or canisters or cartridges. Some APRs are designed for specific hazards such as *M. tuberculosis* (29 C.F.R. 1910.139). Some chemical cartridge respirators are designed to remove specific gases, such as ammonia. All of these filter masks can only be worn by a user if the concentration of contaminant in the air is known and the APR has a high enough fit factor as demonstrated by qualitative fit testing (QLFT) or quantitative fit testing (QNFT) of that individual. The oxygen concentration in the room must also be equal to or greater than 19.5%.

Supplied-air respirators (SAR) have an advantage over APRs because the breathing gas is being supplied from an independent source rather than being filtered. If the SAR maintains the breathing gas at a slightly higher pressure than the ambient air, the SAR is considered positive pressure or pressure demand. This minimizes the opportunity for toxins to be inhaled during inspiration. SARs may not be used in atmospheres immediately dangerous to life and health without an escape air supply. This permits the worker to egress if the flow of breathing gas is interrupted for any reason. One of the most desirable combinations for long-duration use is a SAR used in conjunction with a 30-minute, pressure-demand SCBA. This enables the worker to spend a reasonable time in the contaminated area (hot zone) and still have ample time to decontaminate without fear of exhausting the breathing gas supply. The only disadvantage to this combination is the potential for tangling the supplied air hose and trapping the worker.

The self-contained breathing apparatus affords the highest level of protection. It consists of a high-pressure, DOT-regulated cylinder (typical cylinder pressures range from 2216 psig. to 4500 psig). Cylinder capacities are 45 $ft^3$ for 30-minute units and approximately 90 $ft^3$ for 60-minute units. The high-pressure gas is reduced to very low pressure — 2 to 3.5 in. of water column maximum — by a series of regulators designed to maintain this slight pressure and never allow the pressure in the facepiece to drop lower than the atmospheric pressure outside the facepiece. In this manner, if the facepiece leaks slightly, breathing gas leaks out rather than toxins leaking into the facepiece. All approved respirators employ an end-of-service-life warning device which must sound when 20–25% of the pressure remains in the cylinder. This is designed to allow the wearer to egress to a safe atmosphere. The units also feature pressure gauges that enable the wearer to check the pressure remaining at any time.

## *Open-circuit vs. closed-circuit SCBA*

Although most SCBA in use today is of the open-circuit type described in the preceding paragraph, another type of SCBA, the closed-circuit type (CCSCBA), has enjoyed considerable use for many years for underground mine rescue use. Like the open-circuit units, the CCSCBA employs a high-pressure cylinder of breathing gas. In the CCSCBA, however, the breathing gas is not Grade D breathing air; it is medical-grade oxygen. Instead of each breath coming straight from the high-pressure cylinder, the user's exhaled breath is passed through an absorbent canister filled with a material that absorbs the carbon dioxide from the exhaled air. The remaining oxygen flows into a breathing reservoir where it is available for the user's next breath. Only the amount of oxygen actually used by the body must be replaced from the cylinder. This allows a relatively small cylinder of breathing gas to last a comparatively long time. Mine-rescue teams have used 2- and 4-hour CCSCBA for decades, and the units weigh less than 34 pounds. The ability to produce positive-pressure CCSCBA emerged in the early 1980s, and today some of these units are approved for use in IDLH (immediately dangerous to life and health) atmospheres but not for use in structural firefighting. A disadvantage of the CCSCBA is that each time the gas cylinder is changed, the $CO_2$ absorber must be changed, and it must be stored in a sealed manner to protect the absorbent material from the $CO_2$ that exists in ambient air (approximately 0.05%). When cleaning the respirator, the breathing chamber and hoses must be cleaned and disinfected in addition to the facepiece. Even with all these limitations, for certain long-duration applications where a tethered air-line supply is impractical or dangerous, this unit may be the best choice. Another variation of the CCSCBA involves the use of liquid oxygen instead of a high-pressure cylinder. These units have also been used by mine-rescue teams.

## *Some thoughts on training respirator users*

Although 29 C.F.R. Part 1910.134(c), below, addresses specific training requirements, additional information is provided here to assist the program administrator/instructor. After determining that the individuals who will be trained are physically and psychologically fit to wear the respirator(s) required for the task, the instructor must take into account several factors when designing the training. First is the level of experience/exposure to respirators. If a person was able to hide a claustrophobic reaction to the facepiece during the initial physician screening, this reaction may still surface during the training. If these individuals are allowed to slowly become comfortable with the limited peripheral vision, they may be able to tolerate the respirator. A trainer who is fortunate enough to have individuals in the group who have experience and are comfortable with the respirators can spend more time working with those who are less than excited about wearing respirators.

There is no substitute for knowledge about the hazard, the respirator, and the limitations of the user and the respirator. People who understand the dangers of the hazardous agent and the sealing characteristics of the facepiece rarely try to wear beards, sideburns, and glasses with temple bars that impede the seal of the mask. Quantitative fit testing will resolve any questions about the ability to develop an effective seal with beards, etc. Give these new users detailed information about the hazards and the long-term effects of exposure to the hazards. Make sure they understand operation of the end-of-service-life indicator for both APR filters/canisters and

SCBA service alarms. As with any aspect of safety, frontline supervisors' performance evaluations must include their ability to motivate employees to work safely. The instructor should detail the company's policy, which does not tolerate failure to use and maintain all PPE, especially respirators. Teaching new people to wear respirators comfortably, particularly in hazardous atmospheres, is similar to a test pilot constantly pushing the edges of the aircraft envelope. In this case, the instructor slowly and methodically pushes the respirator user to the edge of their comfort level with the unit. If the instructor is patient and the user is willing, the user's comfort level can continue to expand until the limitations of the respirator have been reached. An instructor who pushes the student too far or too quickly runs the risk of scaring the person, who may refuse to wear the respirator again. In that case, the time and effort involved to that point will have been wasted. The best approach is a training regimen that starts with respirator familiarity and progresses to more difficult and strenuous tasks as the student's confidence rises. Do not expect the entire response team or brigade to be at the same point in their abilities or confidence level. Physical fitness is imperative for maximum performance in emergency situations. A program that helps the employees maintain physical fitness is worth the effort.

Elaborate facilities for training are not essential for all drills. Covering facepieces with wax paper and rubber bands enables the students to learn to communicate without seeing. Tables and chairs can be turned over to build a makeshift maze. Covering the student facepieces also allows the instructor to observe and correct mistakes quickly before they become bad habits. Performing searches in large industrial and commercial facilities is much more difficult than routine searches of small rooms and residential units. Pre-planning exactly how your response team or your local fire department would search your facility is part of good emergency planning. Infrared imaging devices may be necessary and special techniques may have to be developed to find trapped workers or reach the seat of the fire or spill. Remember the five P's:

Pre-
Planning
Prevents
Poor
Performance

Finally, do not try to reinvent the wheel and train your workers by yourself. The industrial hygiene staff, the local fire department instructors, and fellow safety specialists may already have plans and programs that can be adapted to suit the facility. The National Fire Protection Association, the International Society of Fire Service Instructors, the International Fire Service Training Association, and other publishers all offer many excellent publications that can help with respirator training.

## *Elements of a written respirator program*

### *1910.134(c)*

(c) *Respiratory protection program.* This paragraph requires the employer to develop and implement a written respiratory protection program with required worksite-specific procedures and elements for required respirator use. The program must be administered by a suitably trained program administrator. In addition, certain program elements may be required for voluntary use to prevent potential hazards associated with the use of the respirator. The *Small Entity Compliance Guide* contains criteria for the selection of a program administrator and a sample program that meets the requirements of this paragraph. Copies of the *Small Entity Compliance Guide* are available from the Occupational Safety and Health Administration's Office of Publications, Room N 3101, 200 Constitution Avenue, NW, Washington, D.C., 20210 (202-219-4667). You can also download the guide from their website at http://www.osha-slc.gov/SLTC/respiratory_advisor/oshfiles/secgrev.

(c)(1) In any workplace where respirators are necessary to protect the health of the employee or whenever respirators are required by the employer, the employer shall establish and implement a written respiratory protection program with worksite-specific procedures. The program shall be updated as necessary to reflect those changes in workplace conditions that affect respirator use. The employer shall include in the program the following provisions of this section, as applicable:

(c)(1)(i) Procedures for selecting respirators for use in the workplace;

(c)(1)(ii) Medical evaluations of employees required to use respirators;

(c)(1)(iii) Fit testing procedures for tight-fitting respirators;

(c)(1)(iv) Procedures for proper use of respirators in routine and reasonably foreseeable emergency situations;

(c)(1)(v) Procedures and schedules for cleaning, disinfecting, storing, inspecting, repairing, discarding, and otherwise maintaining respirators;

(c)(1)(vi) Procedures to ensure adequate air quality, quantity, and flow of breathing air for atmosphere-supplying respirators;

(c)(1)(vii) Training of employees in the respiratory hazards to which they are potentially exposed during routine and emergency situations;

(c)(1)(viii) Training of employees in the proper use of respirators, including putting on and removing them, any limitations on their use, and their maintenance; and

(c)(1)(ix) Procedures for regularly evaluating the effectiveness of the program.

(c)(2) Where respirator use is not required:

(c)(2)(i) An employer may provide respirators at the request of employees or permit employees to use their own respirators, if the employer determines that such respirator use will not in itself create a hazard. If the employer determines that any voluntary respirator use is permissible, the employer shall provide the respirator users with the information contained in Appendix D to this section ("Information for Employees Using Respirators When Not Required Under the Standard"); and

(c)(2)(ii) In addition, the employer must establish and implement those elements of a written respiratory protection program necessary to ensure that any employee using a respirator voluntarily is medically able to use that respirator, and that the respirator is cleaned, stored, and maintained so that its use does not present a health hazard to the user. Exception: Employers are not required to include in a written respiratory protection program those employees whose only use of respirators involves the voluntary use of filtering facepieces (dust masks).

(c)(3) The employer shall designate a program administrator who is qualified by appropriate training or experience that is commensurate with the complexity of the program to administer or oversee the respiratory protection program and conduct the required evaluations of program effectiveness.

(c)(4) The employer shall provide respirators, training, and medical evaluations at no cost to the employee.

## *1910.134(l)*

(l) *Program evaluation.* This section requires the employer to conduct evaluations of the workplace to ensure that the written respiratory protection program is being properly implemented, and to consult employees to ensure that they are using the respirators properly.

(l)(1) The employer shall conduct evaluations of the workplace as necessary to ensure that the provisions of the current written program are being effectively implemented and that it continues to be effective.

(l)(2) The employer shall regularly consult employees required to use respirators to assess the employees' views on program effectiveness and to identify any problems. Any problems that are identified during this assessment shall be corrected. Factors to be assessed include, but are not limited to:

(l)(2)(i) Respirator fit (including the ability to use the respirator without interfering with effective workplace performance);

(l)(2)(ii) Appropriate respirator selection for the hazards to which the employee is exposed;

(l)(2)(iii) Proper respirator use under the workplace conditions the employee encounters; and

(l)(2)(iv) Proper respirator maintenance.

## *1910.134(m)*

(m) *Recordkeeping*. This section requires the employer to establish and retain written information regarding medical evaluations, fit testing, and the respirator program. This information will facilitate employee involvement in the respirator program, assist the employer in auditing the adequacy of the program, and provide a record for compliance determinations by OSHA.

(m)(1) *Medical evaluation*. Records of medical evaluations required by this section must be retained and made available in accordance with 29 C.F.R. 1910.1020.

(m)(2) *Fit testing*.

(m)(2)(i) The employer shall establish a record of the qualitative and quantitative fit tests administered to an employee including:

(m)(2)(i)(A) The name or identification of the employee tested;

(m)(2)(i)(B) Type of fit test performed;

(m)(2)(i)(C) Specific make, model, style, and size of respirator tested;

(m)(2)(i)(D) Date of test; and

(m)(2)(i)(E) The pass/fail results for QLFTs or the fit factor and strip-chart recording or other recording of the test results for QNFTs.

(m)(2)(ii) Fit test records shall be retained for respirator users until the next fit test is administered.

(m)(3) A written copy of the current respirator program shall be retained by the employer.

(m)(4) Written materials required to be retained under this paragraph shall be made available upon request to affected employees and to the Assistant Secretary or designee for examination and copying.

# *chapter four*

# *Fire and explosion hazards*

> *"He is most free from danger, who, even when safe, is on his guard."*
>
> —Publilius Syrus

> *"The quality of a person's life is in direct proportion to their commitment to excellence, regardless of their chosen field of endeavor."*
>
> —Vince Lombardi

According to the Federal Emergency Management Agency and The U.S. Fire Administration, America has one of the highest fire death rates of any of the industrialized nations in the world. The following statistics published by the National Fire Data Center illustrate the size and scope of the problem.

## *Overall fire picture**

- In 1998, the U.S. fire death rate was 14.9 deaths per million population.
- Between 1994 and 1998, an average of 4400 Americans lost their lives and another 25,100 were injured annually as the result of fire.
- About 100 firefighters are killed each year in duty-related incidents.
- Each year, fire kills more Americans than all natural disasters combined.
- Fire is the third leading cause of accidental death in the home; at least 80% of all fire deaths occur in residences.
- About 2 million fires are reported each year. Many others go unreported, causing additional injuries and property loss.
- Direct property loss due to fires is estimated at $8.6 billion annually.

* http://www.usfa.fema.gov/nfdc/statistics.htm.

## *National fire problem*

The U.S. has a severe fire problem, more so than is generally perceived. Nationally, there are millions of fires, thousands of deaths, tens of thousands of injuries, and billions of dollars lost — which makes the U.S. fire problem one of great national importance. The following table and text illustrate the number of, and trends in, fires, deaths, injuries, and dollar losses in the U.S. from 1989 to 1998.*

| Year | No. of Fires | No. of Deaths | No. of. Injuries | Direct Dollar Loss (millions) |
|---|---|---|---|---|
| 1989 | 2,115,000 | 5410 | 28,250 | $10,951 |
| 1990 | 2,019,000 | 5195 | 28,600 | $9385 |
| 1991 | 2,041,500 | 4465 | 29,375 | $10,906 |
| 1992 | 1,964,500 | 4730 | 28,700 | $9276 |
| 1993 | 1,952,500 | 4635 | 30,475 | $9279 |
| 1994 | 2,054,500 | 4275 | 27,250 | $8630 |
| 1995 | 1,965,500 | 4585 | 25,775 | $9182 |
| 1996 | 1,975,000 | 4990 | 25,550 | $9406 |
| 1997 | 1,795,000 | 4050 | 23,750 | $8525 |
| 1998 | 1,755,500 | 4035 | 23,100 | $8629 |

### *Where fires occur*

- There were 1,755,500 fires in the U.S. in 1998; of these:
  - 41% were outside fires.
  - 29% were structure fires.
  - 22% were vehicle fires.
  - 8% were fires of other types.
- Residential fires represent 22% of all fires and 74% of structure fires.
- Fires in 1- and 2-family dwellings most often start in the:
  - Kitchen (23.5%)
  - Bedroom (12.7%)
  - Living room (7.9%)
  - Chimney (7.1%)
  - Laundry area (4.7%)
- Apartment fires most often start in the:
  - Kitchen (46.1%)
  - Bedroom (12.3%)
  - Living room (6.2%)
  - Laundry area (3.3%)
  - Bathroom (2.4%)
- The South has the highest fire death rate per capita, with 18.4 civilian deaths per million population.

* http://www.usfa.fema.gov/nfdc/national.htm; *Fire Loss in the United States 1987–1996,* 11th ed., National Fire Protection Association, Quincy, MA, 1998.

- Of all fatalities, 80% occur in the home. Of these, approximately 85% occur in single-family homes and duplexes.

### *Causes of fires and fire deaths*

- Cooking is the leading cause of both U.S. home fires and home fire injuries. Cooking fires often result from unattended cooking and human error, rather than mechanical failure of stoves or ovens.
- Careless smoking is the leading cause of fire deaths. Smoke alarms and smolder-resistant bedding and upholstered furniture are significant fire deterrents.
- Heating is the second leading cause of both residential fires and fire deaths; however, heating fires are a greater problem in single-family homes than in apartments. Unlike apartments, the heating systems in single-family homes are often not professionally maintained.
- Arson is the third leading cause of both residential fires and residential fire deaths. In commercial properties, arson is the major cause of deaths, injuries, and dollar loss.

### *Who is most at risk*

- Senior citizens age 70 and over and children under the age of 5 have the greatest risk of fire death.
- The fire death risk among seniors is more than double the average population.
- The fire death risk for children under age 5 is nearly double the risk of the average population.
- Children under the age of 10 accounted for an estimated 17% of all fire deaths in 1996.
- Men die or are injured in fires almost twice as often as women.
- African-Americans and American Indians have significantly higher death rates per capita than the national average.
- Although African-Americans comprise 13% of the population, they account for 26% of fire deaths.

### *What saves lives*

- A working smoke alarm dramatically increases a person's chance of surviving a fire.
- Approximately 88% of U.S. homes have at least one smoke alarm; however, these alarms are not always properly maintained and as a result might not work in an emergency. There has been a disturbing increase over the last 10 years in the number of fires that occur in homes with nonfunctioning alarms.
- It is estimated that over 40% of residential fires and three fifths of residential fatalities occur in homes with no smoke alarms.
- Residential sprinklers have become more cost effective for homes. Currently, few homes are protected by them.

## *Nonresidential properties*

Much of the effort in fire prevention, both public and private, has gone into protecting nonresidential structures, and the results have been highly effective, especially when compared to the residential fire problem. Between 1994 and 1998, nonresidential structures represented approximately 4% of fire deaths, 10% of fire injuries, 31% of total fire dollar loss, and 8% of all fires. The following table and text illustrate the number of, and trends in, fires, deaths, injuries, and dollar losses that occurred in nonresidential properties from 1989 to 1998.

| Year | No. of Fires | No. of Deaths | No. of Injuries | Direct Dollar Loss (millions) |
|---|---|---|---|---|
| 1989 | 174,500 | 220 | 3275 | $4326 |
| 1990 | 157,000 | 285 | 3425 | $2868 |
| 1991 | 162,500 | 190 | 3125 | $3097 |
| 1992 | 165,500 | 175 | 2725 | $3342 |
| 1993 | 151,500 | 155 | 3950 | $2703 |
| 1994 | 151,500 | 125 | 3100 | $2636 |
| 1995 | 148,000 | 290[a] | 2600 | $3257 |
| 1996 | 150,500 | 140 | 2575 | $2971 |
| 1997 | 145,500 | 120 | 2600 | $2502 |
| 1998 | 136,000 | 170 | 2250 | $2326 |

[a] This figure reflects 168 civilian deaths that occurred in the explosion and fire at the federal office building in Oklahoma City on April 19, 1995.

*Source:* http://www.usfa.fema.gov/nfdc/non-resident.htm; Fire Loss in the United States 1987–1996, 11th ed., National Fire Protection Association, Quincy, MA, 1998.

## *Dealing with the fire problem*

To deal effectively with the fire problem, a strategy must be developed. The following steps should be taken to minimize the negative impacts of hostile fires:

1. Assess the potential for fire.
2. Maximize fire prevention efforts.
3. Utilize early detection methods.
4. Utilize automatic suppression systems.
5. Slow the growth of potential fires.
6. Prepare a fire pre-plan and train employees in it to minimize risk and maximize safety and property protection.

### *Step one: assess your potential for fires and explosions*

Take a thorough walk through the facility to determine how much combustible material you have and the proximity of those combustibles to ignition sources. At the same time, try to determine what natural or artificial boundaries exist due to building construction and compartmentalization. Are materials with tremendous heat release and ease of ignition isolated from potential ignition sources? If building compartments exist, are they fire rated with suitable fire-rated closures to protect all openings?

Does your facility have any unique fire protection problems that require special extinguishing systems or special extinguishing agents? If these systems exist, have they been inspected in accordance with the appropriate section of the National Fire Protection Association (NFPA) Code? If your facility is protected by an automatic sprinkler system, was it designed, installed, inspected, and maintained based upon the current fire load and occupancy use? Are there any areas of the facility that are not protected by automatic sprinklers? Are the private and public water-supply systems adequate and maintained to ensure reliability and dependability? Have the public and private hydrants been flowed and maintained regularly? Are all control valves on all fire protection equipment of the indicating type? Are all the control valves electronically supervised and/or chained and locked in the open position?

Contact your insurance carrier to ask for the insurance service office (ISO) rating of all the fire departments near your facility. The ISO rating schedule and system are used to provide information for insurance companies regarding a particular fire department's ability to extinguish fires. Class 1 indicates the highest level of protection, while Class 10 reflects little or no capability. Use this information as a starting point to evaluate the ability of local response agencies to help you in the event of a fire. What is their response time to your facility? Most ISO Class 1 fire departments will have sufficient staffing to consistently arrive at your facility within a prescribed time based upon their distance from your facility. Many communities in America are protected by volunteer fire departments whose response time and staffing level may vary depending upon the time of day. Invite the local fire department in for a tour of your facility. Ask them questions to determine how knowledgeable they are about the facility and, more important, how eager they are to learn and accept the challenges of protecting your facility. The bottom line here is that if you have a solid, Class 1 fire department located across the street, you may only have to provide them with technical support and technical assistance in the event of a fire. However, if you have a Class 9 or 10 fire department with a 15-minute or longer response time, you will have to rely on the facility's automatic fire protection and whatever level of manual suppression force (fire brigade) you can assemble, train, and equip. All too often company risk managers will tell new safety managers, "You never want to start a fire brigade because they are too expensive." The decision to establish a fire brigade cannot be made in a vacuum. If you are faced with long response times and/or questionably trained and equipped response agencies, a fire brigade may be considerably cheaper than having to rebuild the facility and attempt to regain the customer base lost while your facility was out of business.

### *Step two: maximize fire prevention efforts*

To the extent practical and possible, try to keep the items that can burn away from the sources that can ignite them. This is especially true when dealing with materials that are easily ignited or have an unusually high fire load or rapid fire growth. The very nature of many industrial operations prohibits complete isolation of combustibles and ignition sources so your only choice is to minimize the exposure by building storage rooms for dangerous commodities and keeping no more than a one-day supply in the area where the ignition sources are present. Consider facility design features such as MFL (maximum foreseeable loss) walls, which limit fire spread, and ensure that the automatic sprinkler protection is adequate for the hazard. Make sure that sprinklers are inspected and maintained according to the appropriate sections of the NFPA fire code. Advice concerning standard sprinklers is found in NFPA 13, but storage occupancy information including sprinkler design can be found in the appropriate NFPA 230 series fire code based upon the material being stored. Educate your people and help them understand the implications of a major fire; for example, some facilities that suffer a major fire may not rebuild in that same location, as the new facility may be moved to a part of the world where labor or energy costs are lower. Hold supervisors accountable for the fire prevention efforts of their employees in their performance evaluations.

### *Step three: utilize early detection methods*

Make sure that your facility is protected throughout with an appropriate fire alarm system. Use the early-warning type of detectors where practical. If explosion hazards exist, the system will have to be fast enough to sense the explosion and react to discharge the extinguishing agent after the explosion starts but before the pressures rise to damaging levels. These exotic systems are custom designed and installed for each hazard they protect against. They are routinely used in conjunction with explosion venting. Advances in laser detectors, multiple sensing detectors, and air sampling systems are designed to provide early detection and warning while minimizing false alarms. Proper inspection, testing, and maintenance will ensure that the system is operational and will minimize false non-fire activations. Consult the NFPA 72 series for information regarding fire alarm system design, installation, inspection, and maintenance.

### *Step four: utilize automatic suppression systems*

Since the 1800s and the advent of the automatic sprinkler head, statistics have proven that sprinklered buildings are much less likely to be destroyed by fire. Insurance companies recognize this and give substantial discounts to facilities that choose to install, inspect, and maintain automatic fire sprinklers. If the occupancy use and/or fire load have increased since the original sprinkler system was installed, new sprinkler heads with larger diameter orifices may be able to increase the density (gallons per minute per square foot). If a hydraulic analysis of the system proves the existing sprinkler piping and water supply will accommodate the larger diameter heads, the system can be upgraded by switching the heads. In other cases, it may be necessary to replace some or all of the piping to upgrade the fire protection system. While NFPA 13 provides guidance for standard sprinkler systems, storage occupancies have additional requirement listed in the individual storage standards of the NFPA 230 series. Many industrial installations have sufficient quantities of flammable liquids present to merit the installation of automatic or manually activated foam-water systems for controlling Class B fires. These must be maintained in accordance with the appropriate NFPA 11 or 11A. If water in sufficient flow and pressure is not available, one or more fire pumps may be necessary. These pumps must be installed, inspected, tested, and maintained under the regulations of NFPA 20.

*Step five: slow the growth of potential fires*

Recognizing that fire prevention efforts will never be 100% successful, it is necessary to make plans in the event that a fire does occur. The National Fire Protection Association lists the following approaches or strategies to limit the growth of fires after they start:*

1. Restrict materials used in contents and furnishings to reduce heat release, reduce the smoke-generation rate, and prevent unusually high amounts of toxic materials relative to the quantity of smoke generated.
2. Add fire retardant to materials to slow growth and heat release.
3. Use fire resistive barriers to slow the spread of fire to large secondary items.
4. Restrict fuel load by limiting contents based on total fuel potential.
5. Restrict linings, wall coverings, ceiling coverings, and floor coverings of rooms to prevent rapid flame spread.
6. Restrict the use of combustible materials in concealed spaces.
7. Require safe handling of large quantities of potential fuel.

*Step six: prepare a fire pre-plan and train employees in it to minimize risk and maximize life safety and property protection*

The Occupational Safety and Health Act gives employers three choices regarding their response to fire emergencies.

1. First, every employer must develop a fire prevention plan and evacuation plan and train employees on how to report fires and sound the evacuation alarm. The employer must also post evacuation routes and train employees on exiting procedures.
2. An employer may also elect to have designated employees perform specific firefighting duties as trained and equipped. These duties could range from fighting only small, incipient-level fires with extinguishers or small hose lines up to interior structural firefighting.

* *Fire Protection Handbook,* 17th ed., National Fire Protection Association, Quincy, MA, 1991.

Again, the training requirements and equipment must be consistent with the expected level of response in the fire brigade mission and operating procedures.

3. The employer may choose to permit all employees to engage in firefighting duties consistent with their training and equipment. This option is usually reserved for incipient-stage brigades, and the employees are responsible for using extinguishers or small hose lines to extinguish small fires in their work areas.

In the case of interior structural brigades, the training and equipment are equivalent to their municipal firefighting counterparts. In addition to training the brigade members, the level of training for the brigade officers increases with increased levels of response. If your brigade has employees who also belong to response agencies outside their employment with your company, their training with the outside agencies can count toward required brigade training. The experience these individuals bring to the brigade often enables them to lead, motivate, and train other brigade members.

Occasionally, someone will say, "I don't have to comply with fire brigade training and equipment requirements because our team is a Hazardous Materials Response Team" or some other name. It does not matter what the team is called. If the mission and operating procedures allow them to use fire protection equipment for controlling fires, the team must be trained and equipped consistent with that level of response. Utilizing a combined response team makes good sense. Nearly all emergency response team members must have annual physicals to ensure their ability to perform strenuous work and wear respirators. They must understand how to function under an incident command structure and have at least hazardous materials awareness level training. Many companies combine fire, haz-mat response, emergency medical care, and confined space entry/rescue under a single team with individuals qualified at various levels depending upon their training and abilities. Another often overlooked advantage of spending the time and money to train and equip a response team is the ability of team members to recognize hazardous situations and mitigate them before a disaster occurs.

# *chapter five*

# *Confined space hazards*

> *"It is a good thing to learn caution by the misfortune of others."*
> —Publilius Syrus

> *"All's to be fear'd where all is to be lost."*
> —Lord Byron

For many years, employees were being killed or injured while working in areas not designed for continuous human occupancy. Even worse, more than one half of the fatalities in these spaces were rescuers. In a typical scenario, an employee entered the space to inspect, repair, or perform routine maintenance in the space. Either because the conditions in the space were not checked before entry or because conditions deteriorated after the employee entered the space, the original employee failed to emerge from the space. Coworkers or even emergency response personnel all too often enter the space in an attempt to rescue the worker and become fatalities themselves. Although asphyxiation is one of the primary causes of death, other causes include engulfment, entanglement with moving mechanical equipment, electrocution, fires, explosions, and toxic gases and corrosive liquids. OSHA's response to the problem has been a pro-active approach. 29 C.F.R. 1910.146 is concerned with the dangers of the confined spaces, but this regulation mandates the use of an administrative control (permit) system to eliminate all of the potential hazards a worker could encounter before the worker enters the space. If the permit-required confined space (PRCS) program is operating effectively, it should prevent entry into an unsafe space and require the entrants to leave the space if conditions deteriorate while they are still in that space. The standard also provides additional requirements for equipping, training, and evaluating rescue teams if an emergency should arise in a permit-required confined space. Even if an employer chooses to prohibit employees from entering a PRCS, the employer still must educate employees about the dangers of that PRCS. The employer must also post signs at each PRCS warning employees of the danger and prohibiting them from entering the PRCS. One of the first questions a new safety manager asks when learning of the PRCS standard is, "What is a confined space and are there any in my facility?"

## *Confined spaces and permit-required confined spaces defined*

According to OSHA, a "confined space:"

1. Is large enough and so configured that an employee can bodily enter and perform assigned work.
2. Has limited or restricted means for entry or exit (for example, tanks, vessels, silos, storage bins, hoppers, vaults, and pits are spaces that may have limited means of entry).
3. Is not designed for continuous employee occupancy.

A "permit-required confined space" means a confined space that has one or more of the following characteristics:

1. Contains, or has the potential to contain, a hazardous atmosphere.
2. Contains a material that has the potential for engulfing an entrant.
3. Has an internal configuration such that an entrant could be trapped or asphyxiated by inwardly converging walls or by a floor which slopes downward and tapers to a smaller cross-section.
4. Contains any other recognized serious safety or health hazard.

Although many confined spaces are enclosed, there are also open-top pits and tanks that meet the definition of both a confined space and a PRCS. Consultants are frequently asked to provide a list of all the equipment necessary for confined space entry and rescue teams; however, it is not possible to determine what equipment will be required until each space has been studied in detail to assess all of the hazards and the best countermeasures to prevent injury.

## *Classification of hazards*

### *Respiratory hazards*

Because asphyxiation is one of the leading causes of death and injury in confined spaces, everyone working in or near a PRCS must understand these hazards. OSHA does not permit anyone to enter or occupy a space containing less than 19.5% oxygen. Even though humans may be able to hold their breath for a couple of minutes, a single breath of an atmosphere without oxygen will result in immediate unconsciousness and death if the victim is not removed to fresh air within minutes. OSHA standard 29 C.F.R. 1910.146 requires monitoring of the space to ensure that at least 19.5% oxygen is available prior to and during any confined space entry. Simple asphyxiants such as nitrogen, argon, carbon dioxide, and any other gas or vapor that can accumulate and displace the oxygen can be fatal in a PRCS.

Toxic respiratory hazards could be present even if sufficient oxygen is available. Toxic atmospheres involve substances that even in very low concentrations can kill or injure workers. Often, industrial processes utilize or have toxic by-products. Some common examples include hydrogen sulfide, sulfur dioxide, chlorine, and carbon monoxide. Refer to the Material Safety Data Sheet (MSDS) or other sources to determine the permissible exposure limit (PEL) for the chemical identified. Specific test equipment must be used to analyze all the potential toxins that may be present in a PRCS. The PRCS must be purged until the level of toxins present is within acceptable limits. Personal protection equipment (PPE), including a self-contained breathing apparatus (SCBA), should be worn to minimize the risk of injury should the concentration of toxins rise before the entrants can escape. If the hazard cannot be identified or the concentration of the hazard cannot be quantified, a positive-pressure, self-contained breathing apparatus is the only acceptable choice. If the hazard and concentration can be determined, an appropriate respirator and PPE should be selected to ensure that the entrants never exceed the PEL for that agent.

## *Corrosive hazards*

Corrosive atmospheres have the ability to cause damage to the skin or any other human tissue they contact. Strong acids and alkali materials are used in many industrial processes and are by-products generated in others. These atmospheres also have the ability to damage or destroy personal protective equipment. All PPE selected for use in corrosive atmospheres should be compatible with the materials in the space. Some SCBA manufacturers offer special elastomers for use in chlorine environments.

## *Energy hazards*

Because confined spaces by definition are not designed for continuous human occupancy, they often lack the physical space, adequate lighting, and other creature comforts that humans demand. These same hazards often conceal other deadly hazards such as energized electrical components, high-pressure gas and liquid transmission lines, and pneumatic or hydraulic equipment and components. The only way to protect entrants from these hazards is to educate them about all the hazards in the space and by isolating and de-energizing every component in the space before allowing them to enter.

### Engulfment (contents) hazards

Another type of PRCS hazard that has claimed lives is engulfment. Failure to isolate valves that direct contents into vessels is one of the problems. Unauthorized entry into open-top containers with sloping sides and containing grain or other similar materials has also resulted in fatalities. One of the primary ways to guard against unauthorized entry and improperly compensated risk is through effective education. Every employee must be taught to appreciate the dangers, and they must understand the company policies that will not tolerate unauthorized entry.

### Fire and explosive atmospheric hazards

Flammable atmospheres pose serious fire and explosion threats to PRCS entrants. The permit system requires the space to be monitored before entry and continually to ensure that flammable vapors do not accumulate while the entrants are inside the space. If a reading of 10% or greater of the lower flammable limit of the gas is reached according to the combustible-gas monitoring instrument, the entrants must exit the space. The probe of the monitoring instrument should be located in the space where the vapor is most likely to be located. For example, most petroleum product vapors are heavier than air and will accumulate near the bottom of enclosures.

### Mechanical equipment hazards

Mechanical devices in many PRCSs are designed to mix, chop, blend, and stir products. If the primary energy to drive the devices is not properly locked out, according to 29 C.F.R. 1910.147, and if the energy stored in devices such as accumulators, capacitors, compressed air tanks, etc. is not bled off, equipment may activate and injure or kill the entrants. It is worth noting that a single machine may have several different energy sources. For example, a woodshop surfacing machine may have a 220-volt power circuit for the cutting head motor, a 110-volt circuit to control the motors that position the feed table, and a 12- or 24-volt DC control circuit to operate the machine controls. All three sources must be checked to verify that no energy is present.

### Fall hazards

The Occupational Safety and Health Administration requires all employees to be protected from falling from any surface 4 feet (29 C.F.R. 1910.23) to 6 feet (29 C.F.R. 1926.501) above adjacent surfaces. A PRCS is not exempt from this duty to prevent falls. Each employee on a walking or working surface (horizontal or vertical surface) with an unprotected side or edge that is 6 feet (1.8 meters) or more above a lower level must be protected from falling by the use of guardrail systems, safety net systems, or personal fall arrest systems. If the only access to a PRCS is by lowering the employee on a rope, a personal fall arrest system may be the only practical choice.

### *Environmental condition hazards*

Other factors such as extreme temperature may compound the hazards present in the PRCS. Steam pipes or HVAC pipes may be hot or cold enough to cause burns to the skin, or the temperature in the space may be hot enough to cause fatigue and collapse or cold enough that frostbite of extremities may be a concern. Any condition that has the potential to distract the entrant from focusing on the other safety hazards could contribute to an accident or injury.

Because many PRCSs are dark, tiny spaces, encountering reptiles, rodents, or insects is another potential hazard. In such a situation thermal imagers could possibly be used, as any living animal will give off heat and be visible even in low or no light conditions. Noise and vibration are environmental conditions that complicate work and/or rescue in a PRCS. Whenever possible, machinery near the PRCS should be stopped to reduce excess noise and vibration and facilitate communication.

Psychological hazards of working in a restricted space can adversely affect workers. Individuals who are susceptible to claustrophobia could have problems before or during entry into confined spaces. Employers should attempt to uncover this tendency through training prior to working in actual hazardous conditions. People with minor or moderate claustrophobia anxiety in these working conditions can often be slowly acclimated through training and can learn to overcome the anxiety. Individuals who cannot do so should not be forced to work in a PRCS, because they could create dangerous conditions for all the entrants.

## *Communication problems*

Communication problems compound the already difficult work conditions in many PRCSs. SCBA manufacturers are beginning to integrate communication systems into the confined space SCBA equipment. These hardwire systems bundle communication wires with the supplied air hose of systems that enable the workers or rescuers to wear smaller escape SCBA cylinders while still remaining in the space long enough to accomplish the assigned task (work or rescue). The voice-activated microphones and speakers in these systems allow workers to communicate with each other and with support personnel outside the space. In high noise environments, it may be necessary to wear headsets to facilitate communication.

In the absence of high-technology communication systems, other methods of communicating can be used with varying degrees of success. The use of a rope with a system of coded tugs has proved effective in some situations. If the space is small enough and quiet enough, a worker outside may be able to talk to the worker in the space without assistance. Workers in tanks have developed systems of tapping on the tank to communicate with the PRCS attendant on the outside. The OSHA standard requires the attendant to remain in communication with the entrant.

Entry and exit from a PRCS often presents significant challenges to the entrants and to rescuers trying to remove victims. Whenever possible, entrance to the PRCS should be by means level with the ground or within a few feet above the ground. All too often this is not possible. Each PRCS space must be evaluated to design adequate means for entering and exiting the space. Many companies manufacture devices designed to facilitate lowering and raising entrants and rescuers into a PRCS. Any equipment purchased for PRCS rescue must be rated for lifting humans. Workers who are lowered into spaces must keep retrieval lines on themselves while in the space unless doing so increases the danger.

Physical conditions inside the space may demand that the biggest, strongest employee performs the work or rescues the worker, but the arrangement of the space may only permit entry of the smallest member of the team. Pipes, wires, hoses, and equipment can entangle an entrant, causing frustration at the least and panic at worst. The presence of grease, oil, water, steam, or other products only adds to the difficulty. Although already mentioned under mechanical hazards, entrants may find live electrical wires in many PRCSs. Do not overlook these hazards when evaluating any PRCS. Training for rescue in spaces with the physical hazards mentioned in this section would benefit from the availability of props designed to simulate conditions in the PRCS. Workers can learn without the actual dangers present and the instructor can observe and provide input and assistance.

## *Atmospheric testing and monitoring*

When selecting equipment to test and monitor confined spaces, several factors should be considered. Colormetric style detector tubes may be fine for isolating and determining the identity and concentration of unknown substances, but they are only a snapshot analysis of the space at the time the sample was drawn into the tube. The conditions in the space may have already changed by the time the color change has been interpreted. Equipment specific to the hazardous substances identified should be utilized in a manner that continuously monitors the conditions in the space for the presence of, for example, oxygen, flammable/explosive vapors, or any possible toxic gases. These readings should be recorded on the PRCS permit before entry and periodically during the operation. Audible and visual warnings should announce hazardous conditions even if the attendant monitoring the space is distracted by some other activity. The equipment selected must be simple, reliable, and easily understood and operated by the PRCS entry and rescue team members. Individual sensors in monitoring equipment can fail at any time, so it is important to rely on suppliers who can service your equipment quickly and to have additional spare sensors and monitors available. When a piece of equipment monitoring the PRCS fails, everyone in the space must exit and remain out of the space until functioning equipment is back on the job ensuring their safety.

## Entry into a permit-required confined space

At a minimum, the following positions are required by the PRCS standard.

### Permit-required confined space entry supervisor

This supervisor is responsible for ensuring that all aspects of the permit have been addressed and that all hazards have been controlled and documented on the appropriate portions of the entry permit. The permit shall be signed by the entry supervisor, verifying the accuracy of information on the permit.

### Permit-required confined space entrants

These individuals will have witnessed the posting of, or had the opportunity to see, the permit indicating all the hazards identified and the measures taken to control them. The entrants must know the hazards in the space and be able to recognize signs, symptoms, and consequences of exposure to the hazards. They will wear all PPE required to work safely in the space and communicate with the attendant outside the space. They are also responsible for notifying the attendant of any changes in the space and for exiting as quickly as possible if a prohibited condition is detected or if the attendant orders an evacuation of the space. Whenever possible, the entrants should work in teams of two; however, some PRCSs will not have sufficient room for more than one person to enter.

### Permit-required confined space attendant

This individual must maintain communication with the entrants. The attendant may monitor more than one PRCS if doing so does not diminish the ability to monitor and protect all the entrants in all of the PRCSs that the individual is responsible for monitoring. The attendant duties include:

- Monitoring conditions inside and outside the space to determine if it is safe for entrants to remain in the space
- Warning unauthorized persons to stay away from the PRCS
- Performing non-entry rescues as specified in the employer's rescue procedure
- Summoning rescue and emergency services as soon as the attendant determines that the entrants may need assistance to escape the PRCS

Attendants may only enter as members of confined space rescue teams if they are properly trained and equipped and after they have been relieved of their attendant duties. Attendants may not engage in any activities that reduce their ability to monitor and protect the entrants in the PRCSs that they are responsible for monitoring.

## *Confined space entry rescue*

Statistics reveal that more than one half the fatalities in confined space incidents are individuals involved in attempting to perform rescues. Procedures must be developed and enforced to prohibit rescue by anyone other than those people trained and equipped. In some cases, PRCS rescue may be delegated to outside or off-premise personnel. If the victims are to have any chance for survival, the rescue team response times must be quick (minutes). If outside-agency personnel are used for PRCS rescues, the agency selected should be advised of specific hazards in those spaces.

## *Permit-required confined space rescue priorities*

The incident commander of the PRCS rescue team must quickly asses the scene and categorize the rescue as one of three different priorities.

### *Priority one*

Priority one involves rescue from life-threatening situations from which the entrants must be removed quickly or they will die. The cause of the emergency could be an injury to a worker in the space or a change in the environment, or both. In this scenario, all medical treatment must be delayed until the entrants are removed from the hazard area. This is termed "snatch and drag" by some rescue technicians. It is hoped that the entrants still have their retrieval lines attached and that the attendant has removed one or more of the entrants prior to arrival of the rescue team.

### *Priority two*

In this scenario, the victim is disabled by a non-life-threatening injury or illness while inside the PRCS. Atmospheric monitoring indicates that environmental conditions inside the space are still safe but the victim requires medical assistance and/or transport from the space. Because the urgency is reduced, the rescuers take the time required to treat the patient, then package the patient for transport from the PRCS and finally remove the patient from the space. This may require special equipment and techniques to immobilize the patient and protect them from further injury. The rescue team members must have the medical expertise to effectively perform first aid as required. There may be situations where the illness or injury may dictate some level of urgency but the environmental conditions in the space do not present a risk to the rescuers.

### *Priority three*

In this scenario, victims are found in a PRCS and the initial rescue size-up (assessment of the scene) reveals that anyone remaining in the space at this time could not have survived. This is a difficult determination to make;

however, failure to do so places the rescue team members in a "life-swapping" situation. If the victims in the space have no chance of survival, it makes no sense to risk further lives for body-recovery operations. Remember that over one half the fatalities in PRCS rescues are would-be rescuers. If there is a potential life to be saved, then the situation is not priority three; it is priority one.

## *Permit-required confined space training*

Permit-required confined space training falls into three categories. After successful completion of *PRCS awareness training* by all newly hired employees, the employee or independent contractor will be able to:

1. Recognize PRCSs by the posted signs and by OSHA/company definitions.
2. List the location of the PRCS nearest their work area(s).
3. Be able to accurately describe the PRCS policies concerning permits, approvals, entry, and rescue.

*Confined space entry training* is provided to everyone who may serve in any capacity in a confined space entry. The training includes all elements of the PRCS entry permit system and approval process, types of hazards and dangers, countermeasures, atmospheric monitoring, authorization, and entry team member duties and responsibilities. The training also contains site-specific hazard and permit information for each PRCS identified in the facility. The nature of each PRCS presents different hazards and unique equipment requirements to enter and work safely. For example, some PRCSs require horizontal entry while others require lifting harnesses, ropes, and cranes to facilitate entry and exit.

*PRCS rescue team member training* starts with confined space entry training and builds upon it. In addition to a working knowledge of safe ways to work in all of the PRCSs of the facility, the rescue team members are given additional training as follows:

- They are made aware of common PRCS safety violations and the historical consequence of those violations.
- They receive training in emergency medical care consistent with the level of response and care the team will be expected to provide; first aid and CPR are considered the minimum training for basic life support.
- They receive training on inspection, operation, and maintenance of every piece of PRCS entry and rescue equipment they will have available. This would include ropes, lifting devices, tripods, harnesses, immobilization equipment, self-contained breathing apparatus, atmospheric monitoring equipment, and communication and incident command procedures.

- In some companies, the emergency response organization may utilize some or all of the same people on both the fire brigade and PRCS team. This facilitates suppression of fires to effect the rescues more quickly, but the commander must ensure that, prior to rescue team entry, none of the rescue team members is involved in firefighting operations. PRCS rescue team members must focus on their roles to protect the rescue team. Backup rescue team members must be ready to enter immediately if the primary rescue team has problems and must be removed.
- The rescue team may enter IDLH (immediately dangerous to life and health) atmospheres to perform rescues, so it is imperative that competency-based testing be conducted to demonstrate that every team member has the knowledge, skills, and attitudes necessary to work safely and not endanger other team members. Skill sheets should allow checking-off of all critical steps associated with the various pieces of equipment and procedures.
- Site-specific training on each PRCS will include identification of all hazards associated with each PRCS and the safest way to effect each type of rescue from that particular confined space. The initial training should focus on preparation for the most frequently entered spaces and the PRCSs that have already been involved in dangerous incidents or have the greatest potential for injuries or fatalities. Practice is the key to building team skills and confidence.
- Sometimes training will reveal rescue team members who suffer from claustrophobia. Often this can be dealt with effectively by slowly introducing the person to a controlled, safe, monitored confined space a little at a time and then allowing the person to stay in the space a few more minutes with each subsequent training session. This method has enabled individuals with extreme anxiety to overcome their fears. In the event that a team member cannot overcome these fears, they must not be permitted to serve as primary rescue team entrants or even as backup rescue team members. They can still serve in many roles outside the space, such as incident commander, atmospheric monitor, rigger (person who erects the necessary lifting devices), or hoist man to raise victims or pull them from the PRCS.
- The incident command system is the modular command and control system used to manage all emergency response incidents. Teaching the PRCS team to understand and work under this system brings direction, order, and control to the scene and allows the rescue commander to maintain an effective span of control in the event that additional resources are required. This also enables the team to work effectively with outside-agency personnel if they are called to assist.

## *Elements of a written confined space program*

1. The program must be in writing.
2. It must identify all confined spaces in the facility.

3. The program must develop and implement a written entry permit system.
4. The program must address the techniques for monitoring the confined space atmosphere before and during the entry.
5. Personnel must be selected, trained, and equipped with the necessary equipment for confined space entry and rescue.
6. Protective equipment must be acquired and entry and rescue team members must train with it prior to entry and/or rescue operations.
7. Procedures must be developed and training provided to ensure that attendants and emergency response team members are competent.
8. This program should be reviewed annually and updated every time a PRCS changes or new ones are added. All elements should be included in the evaluation, including training, equipment, the permit system, and hazard identification of all PRCS.

## *Elements of a PRCS entry permit*

1. Date, time, and location of the PRCS to be entered
2. Section for recording the signatures of entrants, attendants, and entry supervisor
3. List of all the hazards in the PRCS being permitted
4. Corresponding list of all the countermeasures taken to eliminate each of the hazards identified with this PRCS
5. Means for verifying that pre-entry conditions are in compliance with minimum acceptable entry conditions
6. Place on the permit for recording initial and periodic monitoring
7. Equipment available at time of entry

# chapter six

# Electrical safety

*"The opportunity to begin again, more intelligently."*
—Henry Ford, Sr.

*"A pilot who sees from afar will not make his boat a wreck."*
—Amenemhet I

In the industrial environment, one of the main killers of employees is exposure to electricity. To address the various electrical hazards in the workplace, the Occupational Safety and Health Administration has developed several standards primarily governing the use of electricity in the industrial environment. In the area of installation and maintenance, the National Electric Code (NEC) is the primary set of regulations. Safety professionals should be aware that the National Electric Code has been adopted, in part, within the OSHA standards and is often adopted by various state and local agencies, especially in the area of building codes and construction codes.

This chapter addresses several potential electrical hazards, as well as general safeguards. Given the nature and importance of the OSHA Control of Hazardous Energy standard and its relationship to electrical energy, this standard will also be discussed.

## Electrical safety

In a recent study, it was determined that 1000 people are electrocuted yearly in the U.S., most from low-voltage current (see table next page). This issue is a major concern for industry, which must provide the proper equipment and education to protect employees from these hazards.

Effects of 60-Hz Current on an Average Human

| Current Values | Effect |
|---|---|
| *Safe current values* | |
| 1 milliampere or less | Causes no sensation; cannot be felt; at threshold of perception |
| 1–8 milliamperes | Sensation of shock; not painful; muscular control temporarily lost |
| *Unsafe current values* | |
| 8–15 milliamperes | Painful shock; can let go at will; muscular control not lost |
| 15–20 milliamperes | Painful shock; cannot let go; muscular control of adjacent muscles lost |
| 20–50 milliamperes | Painful; severe muscular contractions; breathing difficult |
| 100–200 milliamperes | Ventricular fibrillation, a heart condition that results in death |
| 200+ milliamperes | Severe burns; muscular contractions so severe that chest muscles clamp heart and stop it for duration of shock |

## *General rules for operating around electricity*

Following are recommendations for providing a safe environment while working around electricity.

### *Employers*

- Employers must promulgate and enforce a set of safety rules for their employees that is consistent with the conditions at their workplace.
- The employer will use positive pressure to ensure that employees observe the rules.
- Employers will conduct training sessions for employees so they will know proper, safe procedures.
- Access to hazard locations will be restricted to qualified employees.
- Employers must provide employees with easy access to safety equipment as needed.

### *Employees*

- Be alert.
- Be cautious.
- Know your job.
- Observe the rules.
- Inspect.

- Identify.
- Isolate.

At no time should employees perform jobs they have not been trained to do!

## Planning

### General

- All work performed on a facility's electrical system must be documented in writing.
- The electrical department will assist where needed to keep the plant's on-line diagram and panel directories up to date.

### Work plan

- Pre-work discussions before each job should be held by the responsible supervisor with the employees involved.
- When more than one crew is working at the same location or on the same circuit, the supervisors of all crews should meet and exchange information regarding each crew's duties.
- The senior supervisor on the multi-crew job is responsible for the safety of the entire job.

### Work crews

- Work performed on live voltage should only be performed by a qualified, experienced employee.
- No employee should work alone on a circuit energized above 300 volts.
- No employee should work alone on a circuit energized above 30 volts when the location of the work is not within sight of other employees.
- Non-employees should not work on energized circuits even for the purpose of monitoring or troubleshooting, unless specifically authorized to do so.

## Safety equipment

All employees should be properly trained and wear the required personal protective equipment (PPE) at all times when working with or around electricity.

### Additional apparel considerations

- Employees are encouraged to wear natural-fiber clothes to prevent material melting.
- Long-sleeved shirts should be worn any time there is danger of arcing.

- No loose, conductive jewelry is to be worn while working on electrical equipment, including necklaces and bracelets.
- Shoes and/or overshoes should not be relied upon for protection from electrical dangers.
- Contact lenses are not to be worn while working on circuits energized over 120 volts or on any circuit with a capacity over 199 amps.

## *Personal protective equipment*

### *Protective helmets*

- Only protective helmets complying with ANSI Standard Z89.1 at a Class B rating shall be worn on jobs involving electricity. In Ontario, adherence to Canadian Standards Association Standards Z94.1 M1977 complies with the intent of R.R.O. 1980, Reg. 692, S. 84.
- Nothing is to be placed inside the hard-hat.
- No holes, for any reason, are to be bored in the shell of the helmet.
- Helmets should not be transported on the rear window shelf of a car or truck.
- Regardless of use, protective helmets should be replaced every 5 years.

### *Face and eye protection*

- Safety glasses with side shields and complying with ANSI Standard Z81.1 must be worn when working on circuits energized at over 50 volts or with a capacity of over 199 amps.
- A face shield must be worn whenever switching open switches and disconnects and when racking in or out drawout circuit breakers or switches.

## *Rubber goods*

C.F.R. 1910.137 (Electrical Protective Devices) states that all rubber protective equipment for electrical workers shall conform to the following requirements established in the American National Standards Institute Standards:

| Item | Standard |
|---|---|
| Rubber insulating gloves | J6.6-1967 |
| Rubber matting for use around electrical apparatus | J6.7-1935 (R1962) |
| Rubber insulating blankets | J6.4-1970 |
| Rubber insulating hoods | J6.2-1950 (R1962) |
| Rubber insulating line hose | J6.1-1950 (R1962) |
| Rubber insulating sleeves | J6.5-1962 |

### *Testing and inspection*

- Rubber gloves must be tested by a company-selected testing agency at least every 6 months; every 3 months if the gloves are used daily.

- All other rubber goods must be tested at least every 12 months; every 6 months if they are used at least 3 days per week.
- All rubber goods are to be inspected and air tested before each day's use.
- No out-of-date rubber goods will be used.

*Storage*

- All rubber goods should be stored in an approved container when not in use.
- No rubber goods are to be stored near ozone-producing equipment; this excludes storage near arc-producing equipment.
- If rubber goods must be stored flat, nothing should be stored on top of them.

*Use*

- Rubber gloves must always be used with their leather protectors.
- Rubber gloves must be worn whenever working on energized circuits rated over 300 volts, phase-to-phase or phase-to-ground.
- Rubber sleeves must be used any time hot work over 300 volts is performed inside a deep cabinet or at medium voltages.
- Rubber mats should be used only for secondary shock protection and should not be relied upon as the sole protection against shocks.

### Fiberglass and wood equipment

- Equipment made out of fiberglass and wood (i.e., hot-sticks, switch-sticks, and ladders) must be kept clean and free from grease and oil.
- When equipment becomes scratched or the outer coating begins to chip off, it should be refurbished using a method and material approved by the manufacturer.

## Enforcement of the program

An electrical safety program without proper enforcement will most likely not meet its potential; therefore, it is recommended that a strong enforcement policy be included with this program. This is to ensure the safety of all employees working with or around electricity.

## Electrical safety checklist

- Have safety rules been developed for working around electricity?
- Have employees been educated on hazards associated with electricity?
- Are training sessions conducted to train employees on proper work procedures and the use of PPE?
- Are pre-work discussions held to plan course of duties?
- Is work performed on live voltages?

## *Control of hazardous energy (lockout and tagout)*

The lack of control of hazardous energy in the industrial workplace is responsible for approximately 10% of all serious accidents and a substantial percentage of fatalities in the workplace every year. According to the Bureau of Labor Statistics, failure to shut off power while servicing machinery and other equipment is the primary cause of injuries. To address this hazard, OSHA promulgated its Control of Hazardous Energy standard (commonly referred to as the lockout/tagout standard).

In general, personnel and human resources managers have applauded the lockout/tagout standard. This standard, unlike many preceding standards, is "user friendly" and provides the exact sequencing of steps to be followed in order to safeguard employees. The difficulty experienced by many personnel and safety managers is in maintaining compliance with the standard over a period of time.

The most efficient method for safeguarding against accidental activation of machinery is the development of a lockout and tagout program that achieves and maintains compliance with the OSHA standard. By locking out and/or tagging off power sources, unauthorized use of the machine or equipment is prevented.

A lockout is simply the placement of a substantial locking mechanism on a machine on/off switch or electrical circuit to prevent the power supply from being activated while repairs are made. Lockout procedures are especially effective in preventing injuries to maintenance or repair personnel who may be placed in a hazardous situation by sudden and unexpected activation of machinery while repairs are being made. Lockouts apply not only to electrical hazards but also to all other types of energy (including hydraulic, pneumatic, steam, chemical, and vehicular).

Locking out a machine or piece of equipment can be used in conjunction with, or separate from, the tagging procedures permitted under the standard. Tags are required to be a bright, identifiable color and to be marked with appropriate wording, such as "Danger: Do Not Operate" or other similar warnings. Tags are required to be of a durable nature and are to be affixed securely to all lockout locations. Tags must contain the signature(s) of the employees working on the machinery, the date and time, and the department name or number.

There are four stages in the development and management of an effective lockout and tagout program. The first stage is program development and equipment modification. In this stage, the workplace must be analyzed and a written lockout/tagout program developed. Identification of the potential hazards and, if possible, elimination of exposures to potential injuries by machinery or equipment should be attempted. Additionally, appropriate equipment, including, but not limited to, padlocks, tags, T-bars, and other equipment of a substantial nature, should be purchased during this stage.

The second stage involves the education and training of affected personnel. The personnel and human resources manager must ensure that all training and education are well documented. Hands-on training is highly recommended.

The third stage involves effective monitoring and disciplinary action. The responsibility for ensuring that all employees and equipment covered under this standard are performing in a manner that is in compliance rests solely with the employer. Disciplinary procedures and enforcement thereof are essential to ensuring continued compliance.

The fourth and final stage involves program auditing and program reassessment. The effectiveness of the compliance program can be ensured through periodic evaluation. Program deficiencies can be identified and corrective action initiated to correct the deficiencies identified. The lockout and tagout standard requires employers to establish a written program for locking out and tagging out machinery and equipment. The written program normally contains the following elements:

- Steps for shutting down and securing the machinery; a written program normally details the energy sources for each machine in the workplace and how it should be locked or tagged. All sources of hazardous energy must be listed, and the means for releasing or blocking stored energy should be included.
- Procedural steps for applying locks and tags and specifications for their placement on the equipment or apparatus; identification of the responsible person(s) authorized to apply locks and tags is required.
- Appropriate steps and testing of the machine/equipment required after shutdown and lockout to verify that all energy is safely isolated.
- Procedures to be followed and steps to be taken in restarting the equipment after completion of the work.
- Identification of persons trained and authorized to lockout machinery.
- Procedures for when the task requires group lockout; each employee is required to possess an individual lock, and only the person applying that lock should have a key to the lock. This ensures that, as different team members complete their tasks and remove their locks, remaining members are still fully protected from the hazardous energy.
- Procedures for shift and personnel changes to ensure continuity of the lockout/tagout protection, including provisions for the safe, orderly transfer of lockout/tagout devices between on- and off-duty personnel.
- The requirement that, whenever major replacement, repair, renovation, or modification of machines or equipment is performed and whenever new machines or equipment are installed, energy-isolating devices for such machines or equipment must accept lockout devices.

## *Preparing for a lockout or tagout system*

The following procedures must be followed when preparing for a lockout and/or tagout procedure:

- Conduct a survey to locate and identify all energy-isolating devices to be certain that switches, valves, or other energy-isolating devices apply to the equipment to be locked or tagged out.
- Ensure that all energy sources to a specific piece of equipment have been identified. Machines usually have more than one energy source (electrical, mechanical, or other).
- It is highly recommended to maintain a list of each piece of equipment, with the results of the survey included, so that each time the lockout/tagout procedure must be performed the mechanic or operator can consult the list.

### *Sequence of lockout or tagout system procedure*

These steps should be followed when a machine is locked and/or tagged:

1. Notify all affected employees that a lockout or tagout system is going to be utilized and the reason for doing so. The authorized employee must know the type and magnitude of energy that the machine or equipment uses and must understand its hazards.
2. If the machine or equipment is operating, shut it down by the normal stopping procedures (depress stop button, open toggle switch, etc.).
3. Operate the switch, valve, or other energy-isolating devices so that the equipment is isolated from its energy sources. Stored energy (such as that in springs, elevated machine members, rotating flywheels, and hydraulic systems or air, gas, steam, water pressure, etc.) must be dissipated or restrained by methods such as repositioning, blocking, or bleeding down.
4. Lockout and/or tagout the energy-isolating devices with assigned individual locks or tags.
5. After ensuring that no personnel are exposed, and as a check on having disconnected the energy sources, operate the push button or other normal operating controls to make certain the equipment will not operate.
6. The equipment is now locked out or tagged out.

### *Restoring machines and equipment to normal production operations*

These procedures should be followed to restore machines to normal operation (basically, the reverse of the above lockout section):

1. After the servicing or maintenance is complete and the equipment is ready for normal production operations, check the area around the machines or equipment to ensure that no one is exposed.
2. After all tools have been removed from the machine or equipment, guards have been reinstalled, and employees are in the clear, remove all lockout or tagout devices.

3. Operate the energy-isolating devices to restore energy to the machine or equipment.

## *Training*

Effective employee training and education regarding the lockout/tagout program are vital components to achieving compliance with the OSHA standard:

1. All employees must understand the purpose and use of an energy-control procedure and know what a tagout signifies. They must know why a machine is locked or tagged out and what to do when they encounter a tag or lock on a switch or a device they wish to operate. Because any employee may encounter a lockout tag or lockout, everyone must have a general understanding of lockout/tagout safety.
2. Before machinery shutdown, the authorized employee must know the type and magnitude of energy to be isolated and how to control it. Each machine or type of machine should have a written lockout procedure (preferably attached to the machine).
3. Retraining to ensure employee proficiency must take place when an employee is re-assigned to a different area or machine or when written procedures change. Additionally, all new employees must be properly trained regarding the lockout/tagout program. When outside contractors are brought on site, they should also be informed of the company's lockout procedures. The company, in turn, should be aware of the contractor's procedures and ensure that its personnel understand and comply with the outside contractor's energy-control procedures (as long as they comply with the standard).
4. Each employee authorized to perform maintenance should be fully knowledgeable about all hazardous energy related to specific machinery.
5. The proper sequence of locking out should be fully understood.

When a tagout systems is used, employees must be made aware of the following limitations of tags:

- Tags are essentially warning devices affixed to energy-isolating devices; they do not provide the physical restraint on those devices that locks do.
- When a tag is attached to an energy-isolating device, it is not to be removed without authorization of the person responsible for it, and a tag is never to be bypassed, ignored, or otherwise defeated.
- To be effective, tags must be legible and understandable by all authorized employees, all affected employees, and all other employees whose work operations are or may be in the area.
- Tags and their means of attachment must be made of materials able to withstand environmental conditions encountered in the workplace.

- Tags may evoke a false sense of security; their meaning must be understood as part of the overall energy-control program.
- Tags must be securely attached to energy-isolating devices so that they cannot be inadvertently or accidentally detached during use.

A violation of the above OSHA standards poses a substantial risk of great bodily harm or death. Exposure to hazardous chemicals, infected bodily fluids, or harmful energy by employees without the protection afforded under the OSHA standards places the personnel and human resources manager and other company officers in a position of risk for potential civil or criminal sanctions under the OSH Act or state laws in the event of an accident. When developing and managing compliance programs, personnel and human resources managers are advised to ensure compliance with each and every element prescribed in the OSHA standard, completely document compliance with each element of the standard, and properly discipline trained employees who are not complying with the prescribed compliance procedures. Personnel and human resources managers who are unsure as to the requirements of the above standards or any other applicable standard should obtain assistance in order to ensure compliance.

In summation, working with electricity, in any form, is a dangerous situation. Electricity should be handled with care and all appropriate precautions taken at all times. Electricity, or any energy source, cannot tell the difference between individuals or equipment. Electricity is one of our greatest assets, allowing us to accomplish our work; however, it can also be an indiscriminate killer. Proper preparation, training, and precautions can keep electricity in its cage and make it work for you. However, if you do not properly prepare, this potential killer can be instantaneously released, causing severe damage and death. Be prepared … electricity is nothing to play with!

## *chapter seven*

# *Machine guarding*

*"It's better to be always upon your guard, than to suffer once."*
—Latin proverb

*"When a man has not a good reason for doing a thing, he has one good reason for letting it alone."*
—Sir Walter Scott

Each year, incidents involving human contact with moving machine components result in serious injuries and deaths. Machine guarding and related machinery violations continuously rank among the top ten OSHA citations issued. In fact, the Mechanical Power Transmission (29 C.F.R. 1910.219) and Machine Guarding: General Requirements (29 C.F.R. 1910.212) standards accounted for the number six and number seven top OSHA violations for FY 1997, with 3077 and 3050 federal citations issued, respectively. There are as many hazards created by moving machine parts as there are types of machines. Safeguards are essential for protecting workers from needless and preventable injuries. A good rule to remember is that any machine part, function, or process that may cause injury must be safeguarded. Three general areas that must be considered for guarding include:

1. Point of operation
2. Power transmission areas
3. Other related moving parts

For *point of operation,* the basic machine operation considered is the purpose for which the machine was designed. It may be sawing, drilling, bending, shearing, punching, or otherwise modifying the original material to make it more useful. These same operations often require tremendous force to transform the material. Machine guards must be in place to prevent contact with these hazardous contact points. Often, loose clothing can become entangled with rotating parts and pull the victim into the machine before it can be stopped. Lacerations, amputations, and crushing injuries may result, the severity of which can range from minor injuries to fatalities. The need to facilitate different kinds, shapes, and sizes of material sometimes makes it difficult to install suitable guards at the point of operation, in which case other means for isolating a worker from this hazard area must be employed. Light curtains, interference fences, and electric interlocks, as well as techniques that physically separate the worker from the machine or prevent machine operation if the worker is present, can be successful if they are well designed, installed, and maintained.

Hazards associated with the *power transmission areas* of machines may be close to the point of operation or may be located at a distance from it. To gain mechanical advantage, many machines employ pulleys, gears, cams, and levers to increase or decrease speed or torque, to facilitate movement of cutting heads, or to feed the stock to the point of operation. All of these areas have the potential for crushing or amputating body parts if guards are not in place to prevent access to the components. If these parts are not fully enclosed on the machine, they must have suitable guards to prevent contact. Even totally enclosed machines can cause serious injury if maintenance personnel do not lock out all energy sources before opening up the machine to perform maintenance.

Finally, many machines have *other related equipment* that is used to position the material being processed before, during, or after the point of operation. Examples include pre- and post-positioning tables, feed rollers, material-handling belts, rollers, robot arms, etc. When a worker steps into, or places a body part in, the path of these machine components, flesh and bone are usually no match for the strength of the machine.

The basic types of hazardous mechanical motions include:

1. Rotating
2. Reciprocating
3. Transversing

*Rotating* motion (including in-running nip points) is particularly dangerous because it can quickly pull fingers, arms, and bodies into the machine before the person can escape or stop the machine. Any projection on a rotating surface increases the likelihood that even slow-turning shafts can pull a victim into the machine. In many cases, the rotating part may be the saw or drill designed to perform the desired operation. Even machine operators who are aware of the danger and have a long history of successful operation can be injured and/or killed in a single, careless instant. The obvious solution is to design the machine/human interface in a manner that makes it impossible for the machine to operate if all guards are not in place or if a worker is in the immediate area.

*Reciprocating* motion is another common type of machine motion that transforms linear motion into rotating motion or vice versa. The crankshaft is a common example. It transforms the downward power stroke of the piston into a rotating motion. Many types of saws use reciprocating motion to cut in one direction or stroke and to clear the cutting chips from the work on the opposite stroke. This back and forth action can be horizontal or vertical. Although this type of motion is less likely to pull a person into the machine, it is still possible. If a shirtsleeve is caught on the backstroke of a reciprocating saw, the person's arm may be pulled into place to be cut on the next forward stroke. In this type of motion, the cutting tool is either pushed or pulled across or through the surface of the material being processed or the tool is held rigid and the material is moved back and forth. In addition to the hazard posed by the cutting blade(s), a worker could be injured by the moving machine components or the work material. If the machine moves back and forth at a high rate of speed, a worker could be struck many times in succession before escaping the path.

The third type of machine motion is *transversing*, which involves movement in a straight, continuous line. A belt or chain uses this principle to turn adjacent shafts or carry the material along an assembly line. Any stationery support or equipment along the path of the transverse motion is a potential nip or shear point. The intersection of belts and chains with their gears and pulleys are run-in or nip points. Murphy's law, which proposes that anything that can go wrong will, suggests that if it is possible for a finger or larger body part to physically fit in these spaces, then eventually they will if left unguarded.

## Forming process actions

### Cutting

Cutting operations can involve any of the three basic machine operations above. The principal danger is at the point of operation of the cutting tool(s) Hand saws, band saws, milling machines, and drill presses can injure workers who come in contact with the tools or materials ejected from the work surface in the form of flying chips. These cutting chips can strike anywhere on the body, but the eyes, face, head, and skin are most often injured. Administrative controls, engineering controls, and personal protective equipment can be used to prevent and minimize injuries.

### Punching

Punching operations employ the use of power and force to transform material into different shapes, to cut holes in the material, or to trim excess material. A single motion of a press may turn a blank sheet of material into several finished pieces. In other operations, the material may move from one press to another until the finished part emerges from the last press. The size of the hazard increases with the size of the parts processed and the size of the presses. The automobile manufacturing industry uses huge hydraulic presses to transform flat sheet metal into body panels. A careless person who has managed to bypass safety devices and comes in contact with the mating halves of a die during its operation would probably lose whatever body parts the die closed on. This is not meant to indicate that small presses are any less dangerous. Even a tiny press may exert several hundred pounds of pressure. Fingers, hands, and arms can be crushed or amputated by even small presses. Installation or removal of dies during changeover or maintenance are other opportunities for injury. One half of a large press die may weigh several hundred pounds.

### *Shearing*

Shearing operations are similar to punch-press operations except that the dies are replaced with blades that cut the material passing through. Depending upon the material being cut, this operation may involve hundreds of pounds of pressure. Even a paper shear can easily amputate fingers or arms. Such shears can be driven by pneumatic, hydraulic, or even human power, with the use of simple machines such as the lever and cam.

### *Bending*

Bending involves the use of presses, breaks, or tubing benders to reshape the material. The primary difference between punching and bending is that bending does not remove material from the work piece. For example, aluminum-siding contractors transform flat aluminum trim stock into various complex shapes up to 12 feet long using a common aluminum break. Any stamping machine that employs dies to reshape the part is actually bending the part. Another example is a tubing bender that transforms straight tubing into complex shapes such as an automobile exhaust pipe. Computerized tubing benders are loaded with an appropriate length of straight tube. When activated, the computer advances and turns the stock as need to produce the exact shape for which it was programmed. An employee too close to the operation could be struck by the stock or the tubing-bender components. Light curtains, interlocks, personal protective equipment, and education are the keys to safety with bending machines.

## *What happens when the safety devices fail?*

Each type of safety device employed is a human-designed and -developed piece of equipment that is subject to wear and failure. Safety professionals must analyze the impact of various failure modes on the safe operation of the machine. For example, if a light curtain is wired in such a manner that the photocell receiver completes the circuit to the ground when the beam is received, what happens if the receiver is shorted to the ground by a frayed wire? Will the safety device act as if the path is clear even though a human hand may be in the work area?

Failure mode analysis is an important function that must not be overlooked. Fail-safe is the proper way to design and wire safety devices. Employees anxious to operate machines with less effort or at higher production rates can be extremely innovative in finding ways to circumvent safety devices. An astounded safety manager asked an employee adjusting a light curtain from behind the device (between the press dies) if the machine was properly locked out. The employee responded, "No, do you think I'm stupid enough to cycle the machine while I'm in it?" Education is the only way to overcome these issues of incorrectly compensated risk.

Frontline supervisors interact with the employees more than anyone else in the plant does and stand the best chance of ensuring that the employees do not take unnecessary risks. Therefore, one of the best ways to improve worker safety is to hold supervisors accountable for the safety of their workers and to make employee safety performance a significant portion of a supervisor's performance appraisals. Ideally, the supervisor would be positively reinforced with incentives if the workers exhibited safe work practices. These same incentives could be withheld if the supervisor failed to take this portion of the job seriously and allowed unsafe work practices to continue.

## *Machine/equipment maintenance*

Any time maintenance is being performed on a piece of equipment, the machine should be locked out in accordance with 29 C.F.R. 1910.147. Proper procedures for shutdown, lockout, and startup must be followed to prevent injury. Fixed guards should be designed to allow simple maintenance, such as lubrication, to be performed with the guards in place. Point-of-operation guards may have to be designed based upon the size of the work piece and the nature of the process. They should be sturdy and made of noncombustible material.

For discussion purposes, we can classify safeguards as follows:

1. Types of guards
   a. Fixed
   b. Interlocked
   c. Adjustable
   d. Self-adjusting
2. Guard-related devices
   a. Presence-sensing detectors
      (1) Photoelectrical (optical)
      (2) Radiofrequency (capacitance)
      (3) Electromechanical

   b. Pullback
   c. Restraint
   d. Safety controls
      (1) Safety trip control
         - Pressure-sensitive body bar
         - Safety tripod
         - Safety tripwire cable
      (2) Two-hand control
      (3) Two-hand trip
   e. Isolation gates
      (1) Interlocked
      (2) Other
3. Location/distance from hazard
4. Material-handling methods to improve safety for the machine operator
   a. Automatic feed
   b. Semi-automatic feed
   c. Automatic ejection
   d. Semi-automatic ejection
   e. Robot
5. Miscellaneous aids
   a. Awareness barriers
   b. Miscellaneous protective shields
   c. Hand-feeding tools and holding fixtures

An example of a machine-guarding checklist is located in Appendix H. This checklist is taken from OSHA publication #3067, *Concepts and Techniques of Machine Guarding*.

## *Personal protective equipment*

Personal protective equipment (PPE) remains an important component that must not be ignored in injury prevention. Administrative controls are developed and the employees are taught why these rules and procedures were implemented to motivate them to work safely. Next, engineering controls are developed and installed. Machine guards are primary examples of engineering controls designed to isolate the worker from the hazard. The third means of preventing and minimizing injuries is the use of appropriate personal protective equipment. Safety glasses, eye shields, helmets, boots, gloves, and respirators are examples of PPE. Employee involvement in the selection of PPE often increases their willingness to wear it because they tend to select PPE that fits better, is more comfortable, or is more aesthetically pleasing. Even if the PPE selected ends up costing a little more, the employees' willingness to wear it makes it worth the added cost. PPE that is not worn is expensive and protects no one.

# *chapter eight*

# *Fall hazards*

> *"Every man gets a narrower and narrower field of knowledge in which he must be an expert in order to compete with other people. The specialist knows more and more about less and less and finally knows everything about nothing."*
>
> —Konrad Lorenz

> *"The first step towards knowledge is to know that we are ignorant."*
>
> —Richard Cecil

Injuries and fatalities from falls are a serious occupational safety problem in this country. They are the leading cause of fatalities in the construction industry. In 1995, over 1000 construction workers died on the job. Nearly one third of these fatalities were the result of falls. Falls are also responsible for deaths and injuries in general industry, as well. Investigation of these incidents reveals several contributing factors, including:

- Unstable working surfaces
- Misuse of fall-protection equipment
- Human error

Effective countermeasures that have proven successful include:

- Guardrails
- Fall-arrest systems
- Safety nets
- Covers
- Travel-limiting systems

In 29 C.F.R. 1926.500, the Occupational Safety and Health Administration (OSHA) utilizes the following definitions concerning fall-protection and related equipment:

- *Anchorage:* secure point of attachment for lifelines, lanyards, or deceleration devices.
- *Body belt (safety belt):* a strap with means both for securing it about the waist and for attaching it to a lanyard, lifeline, or deceleration device.
- *Body harness:* straps that may be secured about the employee in a manner that will distribute the fall-arrest forces over at least the thighs, pelvis, waist, chest, and shoulders, with means for attaching it to other components of a personal fall-arrest system.
- *Buckle:* any device for holding the body belt or body harness closed around the employee's body.
- *Connector:* device used to couple (connect) parts of the personal fall-arrest system and positioning-device systems together. It may be an independent component of the system, such as a carabiner, or it may be an integral component of part of the system, such as a buckle or D-ring sewn into a body belt or body harness or a snap-hook spliced or sewn to a lanyard or self-retracting lanyard.
- *Controlled access zone (CAZ):* area in which certain work (e.g., overhand bricklaying) may take place without use of guardrail systems, personal fall-arrest systems, or safety-net systems and access to the zone is controlled.
- *Dangerous equipment:* equipment (such as pickling or galvanizing tanks, degreasing units, machinery, electrical equipment, and other units) that, as a result of form or function, may be hazardous to employees who fall onto or into such equipment.
- *Deceleration device:* any mechanism, such as a rope grab, rip-stitch lanyard, specially woven lanyard, tearing or deforming lanyards, automatic self-retracting lifelines/lanyards, etc., that serves to dissipate a substantial amount of energy during a fall arrest or otherwise limit the energy imposed on an employee during fall arrest.
- *Deceleration distance:* additional vertical distance a falling employee travels, excluding lifeline elongation and free fall distance before stopping, from the point at which the deceleration device begins to operate. It is measured as the distance between the location of an employee's body belt or body harness attachment point at the moment of activation (at the onset of fall-arrest forces) of the deceleration device during a fall, and the location of that attachment point after the employee comes to a full stop.
- *Equivalent:* alternative designs, materials, or methods to protect against a hazard which the employer can demonstrate will provide an equal or greater degree of safety for employees than the methods, materials, or designs specified in the standard.

- *Failure:* load refusal, breakage, or separation of component parts. Load refusal is the point at which the ultimate strength is exceeded.
- *Free fall:* act of falling before a personal fall-arrest system begins to apply force to arrest the fall.
- *Free fall distance:* vertical displacement of the fall-arrest attachment point on the employee's body belt or body harness between onset of the fall and just before the system begins to apply force to arrest the fall. This distance excludes deceleration distance and lifeline/lanyard elongation, but includes any deceleration device slide distance or self-retracting lifeline/lanyard extension before they operate and fall-arrest forces occur.
- *Guardrail system:* barrier erected to prevent employees from falling to lower levels.
- *Hole:* gap or void, 2 inches (5.1 cm) or more in its least dimension, in a floor, roof, or other walking/working surface.
- *Infeasible:* being impossible to perform the construction work using a conventional fall-protection system (i.e., guardrail system, safety-net system, or personal fall-arrest system) or it is technologically impossible to use any one of these systems to provide fall protection.
- *Lanyard:* flexible line of rope, wire rope, or strap that generally has a connector at each end for connecting the body belt or body harness to a deceleration device, lifeline, or anchorage.
- *Leading edge:* edge of a floor, roof, or formwork for a floor or other walking/working surface (such as a deck) that changes location as additional floor, roof, decking, or formwork sections are placed, formed, or constructed. A leading edge is considered to be an unprotected side and edge during periods when it is not actively and continuously under construction.
- *Lifeline:* component consisting of a flexible line for connection to an anchorage at one end to hang vertically (vertical lifeline) or for connection to anchorages at both ends to stretch horizontally (horizontal lifeline) and which serves as a means for connecting other components of a personal fall-arrest system to the anchorage.
- *Low-slope roof:* a roof having a slope less than or equal to 4 in 12 (vertical to horizontal).
- *Lower levels:* those areas or surfaces to which an employee can fall. Such areas or surfaces include, but are not limited to, ground levels, floors, platforms, ramps, runways, excavations, pits, tanks, material, water, equipment, structures, or portions thereof.
- *Mechanical equipment:* all motor- or human-propelled wheeled equipment used for roofing work, except wheelbarrows and mop carts.
- *Opening:* gap or void 30 inches (76 cm) or more high and 18 inches (48 cm) or more wide in a wall or partition through which employees can fall to a lower level.
- *Overhand bricklaying and related work:* process of laying bricks and masonry units such that the surface of the wall to be jointed is on the

opposite side of the wall from the mason, requiring the mason to lean over the wall to complete the work. Related work includes mason tending and electrical installation incorporated into the brick wall during the overhand bricklaying process.

- *Personal fall-arrest system:* system used to arrest an employee in a fall from a working level. It consists of an anchorage, connectors, and a body belt or body harness and may include a lanyard, deceleration device, lifeline, or suitable combinations of these. As of January 1, 1998, the use of a body belt for fall arrest is prohibited.
- *Positioning-device system:* body-belt or body-harness system rigged to allow an employee to be supported on an elevated vertical surface, such as a wall, and to work with both hands free while leaning.
- *Rope grab:* deceleration device that travels on a lifeline and automatically, by friction, engages the lifeline and locks so as to arrest the fall of an employee. A rope grab usually employs the principle of inertial locking, cam/level locking, or both.
- *Roof:* exterior surface on the top of a building. This does not include floors or formwork that, because a building has not been completed, temporarily become the top surface of a building.
- *Roofing work:* hoisting, storage, application, and removal of roofing materials and equipment, including related insulation, sheet metal, and vapor barrier work but not including the construction of the roof deck.
- *Safety-monitoring system:* safety system in which a competent person is responsible for recognizing and warning employees of fall hazards.
- *Self-retracting lifeline/lanyard:* a deceleration device containing a drum-wound line that can be slowly extracted from, or retracted onto, the drum under slight tension during normal employee movement and which, after onset of a fall, automatically locks the drum and arrests the fall.
- *Snaphook:* connector comprised of a hook-shaped member with a normally closed keeper, or similar arrangement, that may be opened to permit the hook to receive an object and, when released, automatically closes to retain the object. Snaphooks are generally one of two types: (1) locking type, with a self-closing, self-locking keeper that remains closed and locked until unlocked and pressed open for connection or disconnection; or (2) non-locking type, with a self-closing keeper that remains closed until pressed open for connection or disconnection. As of January 1, 1998, the use of a non-locking snaphook as part of personal fall-arrest systems and positioning-device systems is prohibited.
- *Steep roof:* a roof having a slope greater than 4 in 12 (vertical to horizontal).
- *Toeboard:* low, protective barrier that prevents the fall of materials and equipment to lower levels and provides protection from falls for personnel.
- *Unprotected sides and edges:* any side or edge (except at entrances to points of access) of a walking/working surface (e.g., floor, roof, ramp,

or runway) where there is no wall or guardrail system at least 39 inches (1.0 m) high.

- *Walking/working surface:* any surface, whether horizontal or vertical, on which an employee walks or works, including, but not limited to, floors, roofs, ramps, bridges, runways, formwork, and concrete-reinforcing steel but not including ladders, vehicles, or trailers on which employees must be located in order to perform their job duties.
- *Warning-line system:* barrier erected on a roof to warn employees that they are approaching an unprotected roof side or edge; designates area in which roofing work may not take place without the use of guardrail, body belt, or safety-net systems to protect employees in the area.
- *Work area:* that portion of a walking/working surface where job duties are being performed.

## Fall-protection categories

Fall-protection equipment fits into four functional categories:

1. Fall-arresting
2. Positioning
3. Suspension
4. Retrieval

### Fall-arresting equipment

A fall-arrest system is required if any risk exists that a worker may fall from an elevated position; as a general rule, the fall-arrest system should be used any time a working height of 6 feet or more is reached. Working height is the distance from the walking/working surface to a grade or lower level. A fall-arrest system only comes into service when a fall occurs. A full-body harness with a shock-absorbing lanyard or a retractable lifeline is the only product recommended. A full-body harness distributes the forces throughout the body, and the shock-absorbing lanyard decreases the total fall-arresting forces.

### Positioning

This system holds the worker in place while keeping his hands free to work. When the worker leans back, the system is activated; however, the personal positioning system is *not* specifically designed for fall-arrest purposes.

Effective January 1, 1998, body belts are not an acceptable part of a personal fall-arrest system. A body belt is still acceptable as a positioning device and is regulated by 29 C.F.R. 1926.502(e). The dangers posed by using a body belt as a component in a fall-arrest system include:

- Falling out of the belt
- Serious internal injuries from deceleration forces
- Asphyxiation through prolonged suspension by the belt

### Suspension systems and equipment

This equipment lowers and supports the worker, allowing the worker to use both hands. These systems are widely used in the window-washing and painting industries. Suspension system components are not designed to arrest a free fall and should be used in conjunction with a backup fall-arrest system.

### Retrieval plan

Preplanning for victim retrieval in the event of a fall should be an integral part of the design of a fall-arresting system. This will facilitate quicker access to the victim in the event that medical treatment is required.

## Fall-protection systems

Listed below are different types of fall safety equipment and their recommended usage.

- *Cable positioning lanyard:* designed for corrosive or excess heat environments; must be used in conjunction with shock-absorbing devices.
- *Class 1:* body belts (single or double D-ring) designed to restrain a person in a hazardous work position and to reduce the possibility of falls; they should not be used when fall potential exists, as they are for positioning only.
- *Class 2:* chest harnesses used when there are only limited fall hazards (no vertical free-fall hazard) or for retrieving persons, such as removal of persons from a tank or a bin.
- *Class 3:* full-body harnesses designed to arrest the most severe free falls; best choice of harness for absorbing deceleration forces.
- *Class 4:* suspension belts, which are independent work supports used to suspend a worker such as boatswain's chairs or raising or lowering harnesses.
- *Rail system:* can be used on any fixed ladders as well as curved surfaces as a reliable method of fall prevention.
- *Retractable-lifeline system:* provides fall protection and mobility to user when working at a height or in areas where there is a danger of falling.
- *Rope grab:* deceleration device that travels on a lifeline and is used to safely ascend or descend ladders or sloped surfaces; automatically, by friction, engages the lifeline and locks to arrest a person's fall.
- *Rope lanyard:* offers some elastic properties for all types of arrest; used for restraint purpose.
- *Safety net:* can be used to lessen the fall exposure when temporary floors and scaffolds are not used and the fall distance exceeds 25 feet.
- *Shock absorbers:* when used, the fall-arresting force will be greatly reduced if a fall occurs.
- *Web lanyard:* ideal for restraint purposes where fall hazards are less than 2 feet.

## *Duty to have fall protection*

In general, 29 C.F.R. 1926.501 requires protection from falling for anyone who could fall 6 feet or more. Some notable exceptions are found in various sections of 29 C.F.R. 1926, subpart M, including:

- Requirements relating to roofing work on low-slope roofs, 50 feet (15.25 m) or less in width, allow the use of a safety-monitoring system alone (i.e., without the warning-line system).
- When the employer can demonstrate that it is infeasible or creates a greater hazard to use these systems, the employer shall develop and implement a fall-protection plan that meets the requirements of 29 C.F.R. 1926.502(k).
- The provisions of 29 C.F.R. 1926, subpart M, do not apply when employees are making an inspection, investigation, or assessment of workplace conditions prior to the actual start of construction work or after all construction work has been completed.
- Requirements relating to fall protection for employees working on scaffolds are provided in subpart L of 29 C.F.R. 1926.
- Requirements relating to fall protection for employees working on certain cranes and derricks are provided in subpart N of 29 C.F.R. 1926.
- Requirements relating to fall protection for employees performing steel erection work are provided in 1926.105 and in subpart R of 29 C.F.R. 1926.
- Requirements relating to fall protection for employees working on certain types of equipment used in tunneling operations are provided in subpart S of 29 C.F.R. 1926.
- Requirements relating to fall protection for employees engaged in the construction of electric transmission and distribution lines and equipment are provided in subpart V of 29 C.F.R. 1926.
- Requirements relating to fall protection for employees working on stairways and ladders are provided in subpart X of 29 C.F.R. 1926.

## *Training requirements as set forth in 29 C.F.R. 1926.503*

The employer shall provide a training program for each employee who might be exposed to fall hazards. The program shall enable each employee to recognize the hazards of falling and shall train each employee in the procedures to be followed in order to minimize these hazards. The employer shall ensure that each employee has been trained, as necessary, by a competent person qualified in the following areas:

- Nature of fall hazards in the work area
- Correct procedures for erecting, maintaining, disassembling, and inspecting the fall-protection systems to be used

- Use and operation of guardrail systems, personal fall-arrest systems, safety-net systems, warning-line systems, safety-monitoring systems, controlled-access zones, and other protection to be used
- Role of each employee in the safety-monitoring system when this system is used
- Limitations on the use of mechanical equipment during the performance of roofing work on low-sloped roofs
- Correct procedures for the handling and storage of equipment and materials and the erection of overhead protection
- Role of employees in fall-protection plans
- Standards contained in this subpart

### *Certification of training*

The employer shall verify compliance with paragraph (a) of this section by preparing a written certification record that contains:

- Name or other identity of the employee trained
- Date(s) of the training
- Signature of person who conducted the training or signature of the employer

If the employer relies on training conducted by another employer or completed prior to the effective date of this section, the certification record shall indicate the date the employer determined the prior training was adequate rather than the date of actual training. The latest training certification shall be maintained.

### *Retraining*

When the employer has reason to believe that any affected employee who has already been trained does not have the understanding and skill required by paragraph (a) of this section, the employer shall retrain each such employee. Circumstances where retraining is required include, but are not limited to, situations where:

- Changes in the workplace render previous training obsolete.
- Changes in the types of fall-protection systems or equipment to be used render previous training obsolete.
- Inadequacies in an affected employee's knowledge or use of fall-protection systems or equipment indicate that the employee has not retained the requisite understanding or skill.

# chapter nine

# Hearing protection

*"The trouble with a kitten is*
*That*
*Eventually it becomes a*
*Cat."*
—Ogden Nash

*"Never whisper to the deaf or wink at the blind."*
—Slovenian proverb

In many companies and operations, machinery and activities result in a high decibel noise which can have a permanent impact on the employees working in and around the area over a period of time. To assist companies in protecting their employees from these long-term detrimental health effects, the Occupational Safety and Health Administration (OSHA) developed the Hearing Conservation standard, which can be found at 29 C.F.R. 1910.95. In short, companies possessing noisy areas above the specified level are required to comply with this standard. Companies with no areas or operations above the specific level are not required to comply with the standard.

As noted above, exposure to high decibel noise over a period of time can cause substantial and permanent damage to the hearing of employees. The OSHA standard specifies that, where the sound level exceeds an 8-hour, time-weighted average (TWA) level of 85 decibels, adherence to the standard is required. If required to do so, most companies address the potential of engineering measures to reduce the noise levels, where possible, in the work area before addressing the administrative measures identified in the standard. Where engineering controls cannot reduce the noise levels below 85 decibels, companies are responsible for strict adherence to the OSHA standard, including such requirements as audiometric testing, personal protective equipment, training posting, and other requirements as identified in the standard below.

One of the most important components of any hearing-conservation program is training. With this standard, individual employees who are required to wear hearing protection must be appropriately trained to wear and care for this important equipment. Training required under the Hearing Conservation standard must include the following topics:

- Effect of noise on hearing
- Purpose of hearing protectors
- Advantages, disadvantages, and attenuation rates of various types of hearing protection
- Hearing protection selection, fitting, use, and care
- Purpose of audiometric testing, with an explanation of testing procedures

This important training is usually performed by the safety manager, training officer, or other professional who has acquired the appropriate education. Although supervisors do not normally prescribe the types of protection necessary or provide fitting instructions for hearing conservation protection (i.e., earplugs), it is important that supervisors inspect their employees on a daily basis to ensure that they are properly wearing and caring for their hearing protection. Inspection and monitoring of the wearing of hearing protection by employees can be incorporated into a supervisor's daily safety inspections or can be performed as a separate inspection.

An often overlooked area that is not addressed in this standard is the disposal of the hearing protection after use. Employees should be trained in, and supervisors should ensure, proper disposal and replacement of hearing protection to prevent finding discarded hearing protection in products or in machinery. Any employee not complying with the requirements of the program should be subject to disciplinary action in accordance with company policy.

Unlike other OSHA standards, a written program for hearing conservation is not required; however, most prudent companies and organizations do develop a written hearing-conservation program in order to guide and document their efforts. Additionally, many companies and organizations maintain the required audiometric licensure, training certificates, and calibration documents within their written programs for easy monitoring and access.

In summation, the loss of an individual's hearing can be devastating, and companies with high noise-exposure levels can be hurt by the high cost of claims from long-term employees who have worked in the noise-exposure areas. Prudent companies and organizations should appropriately monitor and assess their work areas to identify potential exposures, if any, and take the required steps to ensure compliance with the OSHA Hearing Conservation standard and to protect their employees from the long-term health effects of noise.

# Hearing conservation*

## 29 C.F.R. 1910.95(a)

Protection against the effects of noise exposure shall be provided when the sound levels exceed those shown in Table G-16 when measured on the A scale of a standard sound-level meter at slow response. When noise levels are determined by octave-band analysis, the equivalent A-weighted sound level may be determined as follows (see Figure G-9).

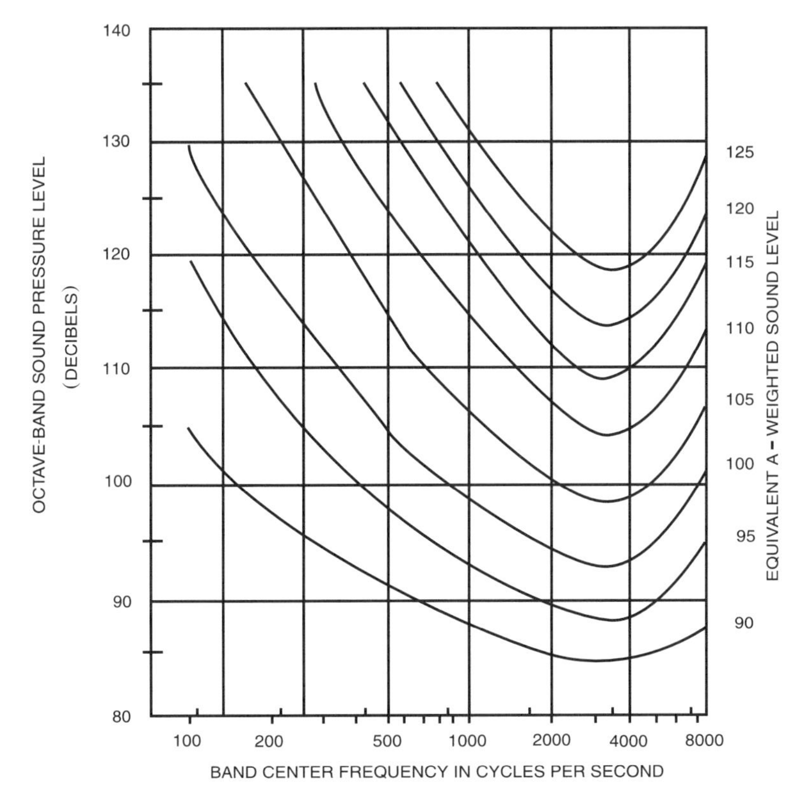


**Figure G-9.** Equivalent A-weighted sound level.

*Equivalent sound level contours.* Octave-band sound-pressure levels may be converted to the equivalent A-weighted sound level by plotting them on the graph shown in Figure G-9 and noting the A-weighted sound level corresponding to the point of highest penetration into the sound level contours. This equivalent A-weighted sound level, which may differ from the actual A-weighted sound level of the noise, is used to determine exposure limits from Table G-16.

* From http:/www.osha-slc.gov/OshStd_toc/OSHA_Std_toc_1910_SUBPART_G.html.

## *29 C.F.R. 1910.95(b)*

(b)(1) When employees are subjected to sound exceeding those listed in Table G-16, feasible administrative or engineering controls shall be utilized. If such controls fail to reduce sound levels within the levels of Table G-16, personal protective equipment shall be provided and used to reduce sound levels within the levels of the table.

***Table G-16.*** Permissible Noise Exposures[a]

| Duration per day (hours) | Sound Level (decibels, slow response) |
|---|---|
| 8 | 90 |
| 6 | 92 |
| 4 | 95 |
| 3 | 97 |
| 2 | 100 |
| 1-1/2 | 102 |
| 1 | 105 |
| 1/2 | 110 |
| 1/4 or less | 115 |

[a] When the daily noise exposure is composed of two or more periods of noise exposure of different levels, their combined effect should be considered, rather than the individual effect of each. If the sum of the following fractions: C(1)/T(1) + C(2)/T(2) + … + C(n)/T(n) exceeds unity, then, the mixed exposure should be considered to exceed the limit value. C(n) indicates the total time of exposure at a specified noise level, and T(n) indicates the total time of exposure permitted at that level. Exposure to impulsive or impact noise should not exceed 140-decibel peak sound-pressure level.

(b)(2) If the variations in noise level involve maxima at intervals of 1 second or less, it is to be considered continuous.

## *29 C.F.R. 1910.95(c): hearing conservation program*

(c)(1) The employer shall administer a continuing, effective, hearing-conservation program, as described in paragraphs (c) through (o) of this section, whenever employee noise exposures equal or exceed an 8-hour, time-weighted average (TWA) sound level of 85 decibels measured on the A scale (slow response) or, equivalently, a dose of 50%. For purposes of the hearing-conservation program, employee noise exposures shall be computed in accordance with Appendix A and Table G-16a, and without regard to any attenuation provided by the use of personal protective equipment.

(c)(2) For purposes of paragraphs (c) through (n) of this section, an 8-hour, time-weighted average of 85 decibels or a dose of 50% shall also be referred to as the action level.

## 29 C.F.R. 1910.95(d): monitoring

(d)(1) When information indicates that any employee's exposure may equal or exceed an 8-hour, time-weighted average of 85 decibels, the employer shall develop and implement a monitoring program.

(d)(1)(i) The sampling strategy shall be designed to identify employees for inclusion in the hearing-conservation program and to enable the proper selection of hearing protectors.

(d)(1)(ii) Where circumstances such as high worker mobility, significant variations in sound level, or a significant component of impulse noise make area monitoring generally inappropriate, the employer shall use representative personal sampling to comply with the monitoring requirements of this paragraph unless the employer can show that area sampling produces equivalent results.

(d)(2)(i) All continuous, intermittent, and impulsive sound levels from 80 decibels to 130 decibels shall be integrated into the noise measurements.

(d)(2)(ii) Instruments used to measure employee noise exposure shall be calibrated to ensure measurement accuracy.

(d)(3) Monitoring shall be repeated whenever a change in production, process, equipment, or controls increases noise exposures to the extent that:

(d)(3)(i) Additional employees may be exposed at or above the action level; or

(d)(3)(ii) The attenuation provided by hearing protectors being used by employees may be rendered inadequate to meet the requirements of paragraph (j) of this section.

## 29 C.F.R. 1910.95(e): employee notification

The employer shall notify each employee exposed at or above an 8-hour, time-weighted average of 85 decibels of the results of the monitoring.

## 29 C.F.R. 1910.95(f): observation of monitoring

The employer shall provide affected employees or their representatives with an opportunity to observe any noise measurements conducted pursuant to this section.

## *29 C.F.R. 1910.95(g): audiometric testing program*

(g)(1) The employer shall establish and maintain an audiometric testing program as provided in this paragraph by making audiometric testing available to all employees whose exposures equal or exceed an 8-hour, time-weighted average of 85 decibels.

(g)(2) The program shall be provided at no cost to employees.

(g)(3) Audiometric tests shall be performed by a licensed or certified audiologist, otolaryngologist, or other physician, or by a technician who is certified by the Council of Accreditation in Occupational Hearing Conservation or who has satisfactorily demonstrated competence in administering audiometric examinations, obtaining valid audiograms, and properly using, maintaining, and checking calibration and proper functioning of the audiometers being used. A technician who operates microprocessor audiometers does not need to be certified. A technician who performs audiometric tests must be responsible to an audiologist, otolaryngologist, or physician.

(g)(4) All audiograms obtained pursuant to this section shall meet the requirements of Appendix C: Audiometric Measuring Instruments.

(g)(5) Baseline audiogram:

(g)(5)(i) Within 6 months of an employee's first exposure at or above the action level, the employer shall establish a valid baseline audiogram against which subsequent audiograms can be compared.

(g)(5)(ii) Mobile test van exception: Where mobile test vans are used to meet the audiometric testing obligation, the employer shall obtain a valid baseline audiogram within 1 year of an employee's first exposure at or above the action level. Where baseline audiograms are obtained more than 6 months after the employee's first exposure at or above the action level, employees shall wear hearing protectors for any period exceeding 6 months after first exposure until the baseline audiogram is obtained.

(g)(5)(iii) Testing to establish a baseline audiogram shall be preceded by at least 14 hours without exposure to workplace noise. Hearing protectors may be used as a substitute for the requirement that baseline audiograms be preceded by 14 hours without exposure to workplace noise.

(g)(5)(iv) The employer shall notify employees of the need to avoid high levels of nonoccupational noise exposure during the 14-hour period immediately preceding the audiometric examination.

(g)(6) Annual audiogram: At least annually after obtaining the baseline audiogram, the employer shall obtain a new audiogram for each employee exposed at or above an 8-hour, time-weighted average of 85 decibels.

(g)(7) Evaluation of audiogram:

(g)(7)(i) Each employee's annual audiogram shall be compared to that employee's baseline audiogram to determine if the audiogram is valid and if a standard threshold shift as defined in paragraph (g)(10) of this section has occurred. This comparison may be done by a technician.

(g)(7)(ii) If the annual audiogram shows that an employee has suffered a standard threshold shift, the employer may obtain a retest within 30 days and consider the results of the retest as the annual audiogram.

(g)(7)(iii) The audiologist, otolaryngologist, or physician shall review problem audiograms and shall determine whether there is a need for further evaluation. The employer shall provide to the person performing this evaluation the following information:

(g)(7)(iii)(A) A copy of the requirements for hearing conservation as set forth in paragraphs (c) through (n) of this section;

(g)(7)(iii)(B) The baseline audiogram and most recent audiogram of the employee to be evaluated;

(g)(7)(iii)(C) Measurements of background sound-pressure levels in the audiometric test room as required in Appendix D: Audiometric Test Rooms.

(g)(7)(iii)(D) Records of audiometer calibrations required by paragraph (h)(5) of this section.

(g)(8) Follow-up procedures:

(g)(8)(i) If a comparison of the annual audiogram to the baseline audiogram indicates that a standard threshold shift as defined in paragraph (g)(10) of this section has occurred, the employee shall be informed of this fact in writing, within 21 days of the determination.

(g)(8)(ii) Unless a physician determines that the standard threshold shift is not work related or aggravated by occupational noise exposure, the employer shall ensure that the following steps are taken when a standard threshold shift occurs:

(g)(8)(ii)(A) Employees not using hearing protectors shall be fitted with hearing protectors, trained in their use and care, and required to use them.

(g)(8)(ii)(B) Employees already using hearing protectors shall be refitted and retrained in the use of hearing protectors and provided with hearing protectors offering greater attenuation if necessary.

(g)(8)(ii)(C) The employee shall be referred for a clinical audiological evaluation or an otological examination, as appropriate, if additional testing is necessary or if the employer suspects that a medical pathology of the ear is caused or aggravated by the wearing of hearing protectors.

(g)(8)(ii)(D) The employee is informed of the need for an otological examination if a medical pathology of the ear that is unrelated to the use of hearing protectors is suspected.

(g)(8)(iii) If subsequent audiometric testing of an employee whose exposure to noise is less than an 8-hour, time-weighted average of 90 decibels indicates that a standard threshold shift is not persistent, the employer:

(g)(8)(iii)(A) Shall inform the employee of the new audiometric interpretation; and

(g)(8)(iii)(B) May discontinue the required use of hearing protectors for that employee.

(g)(9) Revised baseline: An annual audiogram may be substituted for the baseline audiogram when, in the judgment of the audiologist, otolaryngologist, or physician who is evaluating the audiogram:

(g)(9)(i) The standard threshold shift revealed by the audiogram is persistent; or

(g)(9)(ii) The hearing threshold shown in the annual audiogram indicates significant improvement over the baseline audiogram.

(g)(10) Standard·threshold shift:

(g)(10)(i) As used in this section, a standard threshold shift is a change in hearing threshold relative to the baseline audiogram of an average of 10 decibels or more at 2000, 3000, and 4000 Hz in either ear.

(g)(10)(ii) In determining whether a standard threshold shift has occurred, allowance may be made for the contribution of aging (presbycusis) to the change in hearing level by correcting the annual audiogram according to the procedure described in Appendix F: Calculation and Application of Age Correction to Audiograms.

### *29 C.F.R. 1910.95(h): audiometric test requirements*

(h)(1) Audiometric tests shall be pure-tone, air-conduction, hearing-threshold examinations with test frequencies including, as a minimum, 500, 1000, 2000, 3000, 4000, and 6000 Hz. Tests at each frequency shall be taken separately for each ear.

(h)(2) Audiometric tests shall be conducted with audiometers (including microprocessor audiometers) that meet the specifications of, and are maintained and used in accordance with, the American National Standard Specification for Audiometers, S3.6-1969, which is incorporated by reference as specified in Section 1910.6.

(h)(3) Pulsed-tone and self-recording audiometers, if used, shall meet the requirements specified in Appendix C: Audiometric Measuring Instruments.

(h)(4) Audiometric examinations shall be administered in a room meeting the requirements listed in Appendix D: Audiometric Test Rooms.

(h)(5) Audiometer calibration:

(h)(5)(i) The functional operation of the audiometer shall be checked before each day's use by testing a person with known, stable hearing thresholds and by listening to the audiometer's output to make sure that the output is free from distorted or unwanted sounds. Deviations of 10 decibels or greater require an acoustic calibration.

(h)(5)(ii) Audiometer calibration shall be checked acoustically at least annually in accordance with Appendix E: Acoustic Calibration of Audiometers. Test frequencies below 500 Hz and above 6000 Hz may be omitted from this check. Deviations of 15 decibels or greater require an exhaustive calibration.

(h)(5)(iii) An exhaustive calibration shall be performed at least every 2 years in accordance with sections 4.1.2, 4.1.3, 4.1.4.3, 4.2, 4.4.1, 4.4.2, 4.4.3, and 4.5 of the American National Standard Specification for Audiometers, S3.6-1969. Test frequencies below 500 Hz and above 6000 Hz may be omitted from this calibration.

## *29 C.F.R. 1910.95(i): hearing protectors*

(i)(1) Employers shall make hearing protectors available to all employees exposed to an 8-hour, time-weighted average of 85 decibels or greater at no cost to the employees. Hearing protectors shall be replaced as necessary.

(i)(2) Employers shall ensure that hearing protectors are worn:

(i)(2)(i) By an employee who is required by paragraph (b)(1) of this section to wear personal protective equipment; and

(i)(2)(ii) By any employee who is exposed to an 8-hour, time-weighted average of 85 decibels or greater, and who:

(i)(2)(ii)(A) Has not yet had a baseline audiogram established pursuant to paragraph (g)(5)(ii); or

(i)(2)(ii)(B) Has experienced a standard threshold shift.

(i)(3) Employees shall be given the opportunity to select their hearing protectors from a variety of suitable hearing protectors provided by the employer.

(i)(4) The employer shall provide training in the use and care of all hearing protectors provided to employees.

(i)(5) The employer shall ensure proper initial fitting and supervise the correct use of all hearing protectors.

## *29 C.F.R. 1910.95(j): hearing protector attenuation*

(j)(1) The employer shall evaluate hearing protector attenuation for the specific noise environments in which the protector will be used. The employer shall use one of the evaluation methods described in Appendix B: Methods for Estimating the Adequacy of Hearing Protection Attenuation.

(j)(2) Hearing protectors must attenuate employee exposure at least to an 8-hour, time-weighted average of 90 decibels as required by paragraph (b) of this section.

(j)(3) For employees who have experienced a standard threshold shift, hearing protectors must attenuate employee exposure to an 8-hour, time-weighted average of 85 decibels or below.

(j)(4) The adequacy of hearing protector attenuation shall be re-evaluated whenever employee noise exposures increase to the extent that the hearing protectors provided may no longer provide adequate attenuation. The employer shall provide more effective hearing protectors where necessary.

## *29 C.F.R. 1910.95(k): training program*

(k)(1) The employer shall institute a training program for all employees who are exposed to noise at or above an 8-hour, time-weighted average of 85 decibels and shall ensure employee participation in such program.

(k)(2) The training program shall be repeated annually for each employee included in the hearing-conservation program. Information provided in the training program shall be updated to be consistent with changes in protective equipment and work processes.

(k)(3) The employer shall ensure that each employee is informed of the following:

(k)(3)(i) The effects of noise on hearing;

(k)(3)(ii) The purpose of hearing protectors; the advantages, disadvantages, and attenuation of various types; instructions on selection, fitting, use, and care; and

(k)(3)(iii) The purpose of audiometric testing and an explanation of the test procedures.

## *29 C.F.R. 1910.95(l): access to information and training materials*

(l)(1) The employer shall make available to affected employees or their representatives copies of this standard and shall also post a copy in the workplace.

(l)(2) The employer shall provide to affected employees any informational materials pertaining to the standard that are supplied to the employer by the Assistant Secretary.

(l)(3) The employer shall provide, upon request, all materials related to the employer's training and education program pertaining to this standard to the Assistant Secretary and the Director.

## *29 C.F.R. 1910.95(m): recordkeeping*

(m)(1) Exposure measurements: The employer shall maintain an accurate record of all employee exposure measurements required by paragraph (d) of this section.

(m)(2) Audiometric tests:

(m)(2)(i) The employer shall retain all employee audiometric test records obtained pursuant to paragraph (g) of this section:

(m)(2)(ii) This record shall include:

(m)(2)(ii)(A) Name and job classification of the employee;

(m)(2)(ii)(B) Date of the audiogram;

(m)(2)(ii)(C) The examiner's name;

(m)(2)(ii)(D) Date of the last acoustic or exhaustive calibration of the audiometer; and

(m)(2)(ii)(E) Employee's most recent noise exposure assessment.

(m)(2)(ii)(F) The employer shall maintain accurate records of the measurements of the background sound-pressure levels in audiometric test rooms.

(m)(3) Record retention: The employer shall retain records required in this paragraph (m) for at least the following periods:

(m)(3)(i) Noise exposure measurement records shall be retained for two years.

(m)(3)(ii) Audiometric test records shall be retained for the duration of the affected employee's employment.

(m)(4) Access to records: All records required by this section shall be provided upon request to employees, former employees, representatives designated by the individual employee, and the Assistant Secretary. The provisions of 29 C.F.R. 1910.20 (a)–(e) and (g)–(m)(4)(i) apply to access to records under this section.

(m)(5) Transfer of records: If the employer ceases to do business, the employer shall transfer to the successor employer all records required to be maintained by this section, and the successor employer shall retain them for the remainder of the period prescribed in paragraph (m)(3) of this section.

### *29 C.F.R. 1910.95(n): appendices*

(n)(1) Appendices A, B, C, D, and E to this section are incorporated as part of this section and the contents of these appendices are mandatory.

(n)(2) Appendices F and G to this section are informational and are not intended to create any additional obligations not otherwise imposed or to detract from any existing obligations.

### *29 C.F.R. 1910.95(o): exemptions*

Paragraphs (c) through (n) of this section shall not apply to employers engaged in oil and gas well drilling and servicing operations.

### *29 C.F.R. 1910.95(p): startup date*

Baseline audiograms required by paragraph (g) of this section shall be completed by March 1, 1984.

[39 FR 23502, June 27, 1974, as amended at 46 FR 4161, Jan. 16, 1981; 46 FR 62845, Dec. 29, 1981; 48 FR 9776, Mar. 8, 1983; 48 FR 29687, June 28, 1983; 54 FR 24333, June 7, 1989; 61 FR 5507, Feb. 13, 1996; 61 FR 9227, March 7, 1996]

# *chapter ten*

# *Bloodborne pathogens standard*

*"The essence of knowledge is, having it, to use it."*
—Confucius

*" An investment in knowledge always pays the best interest."*
—Benjamin Franklin

## *OSHA bloodborne disease standard program development*

This relatively new Occupational Safety and Health Administration (OSHA) standard was developed to protect employees from possible exposure to HIV, hepatitis, and other bloodborne diseases to which the employee may be exposed in the workplace. For emergency medical services (EMS) organizations, the potential for exposure to bodily fluids (and, thus, bloodborne pathogens) occurs when medical personnel render treatment in homes, accident scenes, or elsewhere. From a legal perspective, providing emergency medical treatment has numerous potential risks for EMS organizations in regard to not only workers' compensation for exposed employees but also tort liability (i.e., negligence), malpractice, and decreased efficacy.

Compliance with the standard is an absolute "must" for EMS organizations. The basic procedure for the development of a program to achieve compliance with this new standard includes:

- Acquire a copy of the current OSHA standard.
- Review this standard with the management team and acquire management commitment and appropriate funding for this program.

- Develop a written program incorporating all required elements of the standard which include, but are not limited to, Universal Precautions, Engineering and Work Practice Controls, personal protective equipment, housekeeping, infectious waste disposal, laundry procedures, training requirements, hepatitis B vaccinations, information to be provided to the physician, medical recordkeeping, signs and labels, and availability of medical records.

Under the scope of universal precautions, the company's operations should be analyzed to provide all necessary safeguards to employees who may have possible contact with human blood. Engineering Controls and Work Practice Controls should be examined and evaluated. Employees should be properly trained in the requirements of this standard. Procedures should be established for the safe disposal of used needles, used personal protective equipment, and other equipment. Such areas as the refrigerator, cabinets, or freezers where blood could be stored should be prohibited for the storage of food or drink (e.g., employees' keeping their lunches in the nurse's refrigerator). Employees must be advised to refrain from eating, drinking, smoking, applying cosmetics or lip balm, or handling contact lenses after possible exposure.

Where there is the potential of exposure, personal protective equipment, such as surgical gloves, gowns, fluid-proof aprons, faceshields, pocket masks, ventilation devices, etc., must be provided to employees. This requirement is especially important for first-aid responders and plant medical personnel who may be exposed when providing first aid to an injured worker. The personal protective equipment must be appropriately located for easy accessibility. Hypoallergenic personal protective equipment should be made available to employees who may be allergic to the normal personal protective equipment.

Employers are required to maintain a clean and sanitary worksite. A *written* schedule for cleaning and sanitizing (disinfecting) all applicable work areas must be implemented and included in the program. All areas, equipment, etc. that have been exposed (e.g., after an accident) must be cleaned and disinfected. Exposed broken glass may *not* be picked up directly with the hands. A dust broom, vacuum, tongs, or other instrument must be used.

All infectious waste, such as bandages, towels, or other items exposed to human blood, must be placed in a leakproof container or bag bearing the appropriate label. This waste must be properly disposed of as medical waste according to federal, state, and local regulations. Please check with the local hospital or governmental agency to learn about the applicable regulations.

Contaminated uniforms, smocks, and other items of personal clothing must be laundered in accordance with the standard. *Do not send contaminated uniforms, smocks, etc. to the in-plant laundry.*

Where required, employees who are at risk of being exposed must undergo a medical examination and acquire a hepatitis B vaccination.

Examples of personnel who may meet this requirement include plant medical personnel and first-aid responders, among others. Training requirements include providing a copy of the OSHA standard and explaining it, as well as discussion of the symptoms and epidemiology of bloodborne diseases; transmission of bloodborne pathogens; methods for recognizing jobs and other activities that could involve exposure to blood; the uses and limitations of practices that will prevent or reduce exposure, such as Engineering Controls and the use of personal protective equipment; the types, proper use, location, removal, handling, decontamination, and/or disposal of personal protective equipment; the basis for selection of personal protective equipment; the hepatitis B vaccination, including information on its efficacy, safety, and benefits; appropriate actions to be taken and persons to contact in an emergency; procedures to be followed if an exposure incident occurs, including the method of reporting, medical follow-up, and medical counseling; and signs and labels. Additional training may be required in applicable laboratory situations and other circumstances.

The employer is required to maintain complete and accurate medical records. These records must include the names and social security numbers of employees; each employee's hepatitis B vaccination record and medical evaluations; results of all physical examinations, medical testing, and follow-up procedures for each employee; physicians' written opinions; and copies of all information provided to the physicians. The employer is charged with the responsibility of maintaining the confidentiality of these records.

The employer is additionally required to maintain all testing records, which must include the dates of all training; names of persons conducting the training; and names of all participants. The training records must be maintained for a minimum of five (5) years. Employees and OSHA must be given the opportunity to review and copy these records upon request.

All containers of infectious waste (including, but not limited to, refrigerators and freezers containing infectious waste, medical disposal containers, and all other containers) *must* be properly labeled. The required label must be fluorescent orange or orange-red, bear the appropriate symbol, and include the lettering "BIOHAZARD" in a contrasting color. This label *must* be affixed to all containers containing infectious waste and must remain affixed until the waste is properly disposed of.

## *Example of a bloodborne pathogens program*

**Bloodborne Pathogens Program**

Developed on ________________. Updated on ________________.

*Contents* *Pages*


Introduction ______
Definitions ______
Exposure control plan: overview ______
Exposure determination ______
Engineering and work practice controls ______
Personal protective equipment ______
Housekeeping ______
Hepatitis B vaccine and vaccination series and post-exposure evaluation ______
Hazard communication ______
Training and education ______
Recordkeeping ______


### *Introduction*

The objective of this manual is to provide EMS, Inc., employees, health professionals, and other managers who are responsible for protecting the health and safety of employees with information they need to comply with Occupational Safety and Health Administration (OSHA) requirements concerning the elimination or minimization of occupationally transmitted bloodborne infections. Included are instructions for compliance with the OSHA Bloodborne Pathogens standard, education and training materials, and guidelines for the medical evaluation and treatment of employees who may be exposed to blood, body fluids, or other materials potentially contaminated with blood or body fluids. It is worth emphasizing that, without meticulous adherence to the concepts and requirements in this manual, the potential for serious injury to health is a very real one; the inherent complexities attendant with eliminating or minimizing potential exposures should not be underestimated.

#### *Instructions*

The manual has been designed to present the Bloodborne Pathogens standard in a readily understandable format and to provide checklists and self-prompting forms that are easy to use and which can assist companies and organizations to achieve compliance. The reader is urged to take a few moments to become familiar with the various sections in this manual. The format of each section consists of text from the standard followed by the recommended forms to be used. Once completed, the forms become the facility's Exposure Control Plan for compliance with the OSHA standard.

*Bloodborne Pathogens Program*

A copy of the written Bloodborne Pathogens Program and Bloodborne Pathogen Training Program for EMS, Inc., as well as the OSHA standard upon which these programs are based, is located in the main office and is readily available for use by all employees. If you should have any questions, please contact your supervisor.

______________________________
*President*

## *Definitions (OSHA defined)*

Please take a moment to read this section carefully. The words and phrases included have meanings specific to the Bloodborne Pathogens standard.

*Blood:* human blood, human blood components, and products made from human blood.

*Bloodborne pathogens:* pathogenic microorganisms that are present in human blood and can cause disease in humans. These pathogens include, but are not limited to, hepatitis B virus (HBV) and human immunodeficiency virus (HIV).

*Clinical laboratory:* a workplace where diagnostic or other screening procedures are performed on blood or other potentially infectious materials.

*Contaminated:* the presence or reasonably anticipated presence of blood or other potentially infectious materials on an item or surface.

*Contaminated laundry:* laundry that has been soiled with blood or other potentially infectious materials or may contain sharps.

*Contaminated sharps:* any contaminated object that can penetrate the skin, including, but not limited to, needles, scalpels, broken glass, broken capillary tubes, and exposed ends of dental wires.

*Decontamination:* the use of physical or chemical means to remove, inactivate, or destroy bloodborne pathogens on a surface or item to the point where they are no longer capable of transmitting infectious particles and the surface or item is rendered safe for handling, use, or disposal.

*Employees at significant risk of exposure:* includes healthcare providers (physicians, nurses, lab technicians), who comprise a risk group because they are routinely exposed to body fluids during the performance of their clinical duties. Those employees who are not part of the primary healthcare team (e.g., security personnel, CPR-trained employees) but as part of their job responsibilities are required to respond to medical emergencies may also be at increased risk due to the greater potential for exposure to blood in emergency situations. Employees who handle medical waste, laundry, medical instruments, or other materials contaminated by body fluids may be at risk of exposure during cleaning, maintenance, or other housekeeping activities. Other medical facility staff such as administrators, managers, administrative assistants, receptionists, etc. are generally not at risk.

*Engineering Controls:* controls (e.g., sharps disposal containers, self-sheathing needles) that isolate or remove the bloodborne pathogens hazard from the workplace.

*Exposure incident:* specific eye, mouth, other mucous membrane, non-intact skin, or parenteral contact with blood or other potentially infectious materials that results from the performance of an employee's duties.

*Handwashing facilities:* a facility providing an adequate supply of running, potable water, soap, and single-use towels or hot-air drying machines.

*Licensed healthcare professional:* a person whose legally permitted scope of practice allows him or her to independently perform the activities required for hepatitis B vaccination and post-exposure evaluation and follow-up.

*HBV:* hepatitis B virus.

*HIV:* human immunodeficiency virus.

*Occupational exposure:* reasonably anticipated skin, eye, mucous membrane, or parenteral contact with blood or other potentially infectious materials that may result from the performance of an employee's duties.

*Other potentially infectious materials:* (1) the following human body fluids: semen, vaginal secretions, cerebrospinal fluids, synovial fluid, pleural fluid, pericardial fluid, peritoneal fluid, amniotic fluid, saliva in dental procedures, any body fluid that is visibly contaminated with blood, and all body fluids in situations where it is difficult or impossible to differentiate between body fluids; (2) any unfixed tissue or organ (other than intact skin) from a human (living or dead); and (3) HIV-containing cell or tissue cultures, organ cultures, HIV- or HBV-containing culture medium or other solutions, blood organs, or other tissues from experimental animals infected with HIV or HBV.

*Parenteral:* piercing mucous membranes or the skin barrier by such events as needle sticks, human bites, cuts, and abrasions.

*Personal protective equipment:* specialized clothing or equipment worn by an employee for protection against a hazard. General workclothes (e.g., uniforms, pants, shirts, or blouses) not intended to function as protection against a hazard are not considered to be personal protective equipment.

*Regulated waste:* liquid or semi-liquid blood or other potentially infectious materials, contaminated items that would release blood or other potentially infectious materials in a liquid or semi-liquid state if compressed, items that are caked with dried blood or other potentially infectious materials and are capable of releasing these materials during handling, contaminated sharps, and pathological and microbiological wastes containing blood or other potentially infectious materials.

*Research laboratory:* a laboratory producing or using research-laboratory-scale amounts of HIV or HBV. Research laboratories may produce high concentrations of HIV or HBV but not in the volume found in production facilities.

*Source individual:* any individual, living or dead, whose blood or other potentially infectious materials may be a source of occupational exposure to the employee. Examples include, but are not limited to, hospital and clinic patients, clients in institutions for the developmentally disabled, trauma victims, clients of drug and alcohol treatment facilities, residents of hospices and nursing homes, human remains, and individuals who donate or sell blood or blood components.

*Sterilize:* the use of a physical or chemical procedure to destroy all microbial life, including highly resistant bacterial endospores.

*Universal Precautions:* an approach to infection control, according to which all human blood and certain body fluids are treated as if known to be infectious for HIV, HBV, and other bloodborne pathogens.

*Work Practice Controls:* controls that reduce the likelihood of exposure by altering the manner in which a task is performed (e.g., prohibiting recapping of needles by a two-handed technique).

## *Exposure Control Plan: overview*

Each company or organization must have a written Exposure Control Plan (ECP). The purpose of the ECP is to eliminate or minimize *"occupational exposures"* (hereafter, Exposures). Following are mandatory elements of the ECP which will be addressed in subsequent sections of this manual.

- Exposure determination (identifying covered employees)
- Schedule and methods of implementation:
  - Methods of compliance
  - Hepatitis B vaccination and post-exposure evaluation
  - Communication of hazards to employees
  - Recordkeeping
- Procedure for evaluation of circumstances surrounding exposure incidents

*Note:* Companies or organizations are responsible for ensuring that the ECP is readily available (during the work shift) for review by covered employees and that a copy is made available to them without charge, within *15 working days* upon request. The ECP must be reviewed *annually* and *whenever necessary* to reflect new or modified tasks and procedures which affect occupational exposure, new employee positions with occupational exposure, and changed employee positions which now include occupational exposure.

## *General information*

Facility: EMS, Inc.

Address: [street address]
[city, state, zip]

### *Bloodborne Pathogens Manual*

Completed by: ______[name] ______________________

Date: ______12/2/99______________________

Date(s) of update reviews: ______scheduled 11/2000 ______________

Update reviewer(s): ______[name(s)]___________________

Review and updating of the appropriate sections in this *Bloodborne Pathogens Manual* will be performed *annually* and whenever there is a need for a change in the exposure determination list (e.g., change in employees who are occupationally exposed; change in tasks, procedures, or jobs that may affect occupational exposure).

Responsible person/party: ______[name] ______________________

A copy of the *Bloodborne Pathogens Manual* will be made readily accessible to all occupationally exposed employees at [location]. Provisions have been made to provide a copy of the Exposure Control Plan within 15 working days of an occupationally exposed employee's request.

Responsible person/party: ______[name] ______________________

### *Schedule of implementation*

| Date Required | ECP Element | Date Completed |
|---|---|---|
| ______________ | __________________ | ______________ |
| ______________ | __________________ | ______________ |
| ______________ | __________________ | ______________ |
| ______________ | __________________ | ______________ |
| ______________ | __________________ | ______________ |

### *Exposure determination*

Each company or organization having employees with Exposure must be identified and included in the Exposure Control Program. This process of identification is known as *exposure determination* and requires the creation of two lists:

1. The first list contains the job classifications in which *all* employees in a particular job classification have Exposure — for example, physicians and nurses, laboratory technicians, and possibly first-aid responders.
2. The second list contains the job classifications in which only *some* of the employees have Exposure, along with the associated tasks or procedures in that job classification in which Exposure may occur — for example, security personnel: performing CPR.

## *Exposure determination list*

**A.1.** Indicate (with an X) those job classifications in which *all* employees may have occupational exposure to bloodborne pathogens:

| | |
|---|---|
| [ X ] Physician | [ X ] First-aid responders |
| [ X ] Nurse | [ X ] Selected management |
| [ X ] Physician's assistant | [ ] ________________ |
| [ X ] Laboratory technician | [ ] ________________ |

*List tasks/procedures associated with occupational exposures:*

First-aid responders: Clean up after accident or spill

______________________________________

______________________________________

**A.2.** Complete evaluations for employees in A.1. (*Note:* All employees listed must be offered full protection under the OSHA standard.)

| | | |
|---|---|---|
| ____________ | ____________ | ____________ |
| ____________ | ____________ | ____________ |
| ____________ | ____________ | ____________ |
| ____________ | ____________ | ____________ |
| ____________ | ____________ | ____________ |

**B.** List job classifications in which *some* (but not all) employees have occupational exposures to bloodborne pathogens (use additional pages as needed).

| Job Classification/ Position | Tasks/Procedures Associated with Occupational Exposures |
|---|---|
| ____________ | ______________________________ |
| ____________ | ______________________________ |
| ____________ | ______________________________ |
| ____________ | ______________________________ |
| ____________ | ______________________________ |
| ____________ | ______________________________ |
| ____________ | ______________________________ |
| ____________ | ______________________________ |
| ____________ | ______________________________ |
| ____________ | ______________________________ |
| ____________ | ______________________________ |
| ____________ | ______________________________ |
| ____________ | ______________________________ |
| ____________ | ______________________________ |

**C.** List employees who may have occupational exposures in the job classifications identified in Section B (use additional pages as needed). (All employees listed must be offered full protection under the OSHA standard).

| Name | Job Classification/Position | Date of Completion |
|---|---|---|
| | | |
| | | |
| | | |
| | | |
| | | |
| | | |
| | | |
| | | |
| | | |
| | | |
| | | |
| | | |
| | | |
| | | |

## *Engineering and Work Practice Controls*

A number of preventive and protective actions must be taken to eliminate or minimize occupational exposures to bloodborne pathogens. This section addresses compliance methods known as Universal Precautions, Engineering Controls, and Work Practice Controls.

### *Universal Precautions*

Universal Precautions is an OSHA-mandated system of infection-control techniques which operates under the premise that the only prudent way of protecting employees from bloodborne pathogens is to treat all body fluids (OSHA-defined) as potentially infectious. Attendant with this philosophy is the necessity to protect all potential portals of entry (e.g., mucous membranes of eyes, nose, mouth; skin) exposed to body fluids in the course of performing job functions (phlebotomy, wound care, PAP smears, etc.).

### *Engineering and Work Practice Controls*

Engineering Controls and Work Practice Controls must be employed as means of reducing occupational exposures. Engineering Controls accomplish this by using specifically designed equipment and facilities, while Work Practice Controls reduce exposure by altering the manner in which a task is performed.

| Engineering Controls | Work Practice Controls |
|---|---|
| Handwashing facilities | Handwashing |
| Sharps | Use of sharps |
| Sharps containers | No recapping of needles |
| Red bags/containers (appropriately labeled cabinets and storage containers) | No eating, drinking, applying cosmetics, or handling contacts in prohibited areas |

*Note:* Each company or organization is responsible for assuring that Engineering Controls are made available, utilized, and maintained on a regular schedule; that employees follow the Work Practice Controls identified for the job or task; and that employees are trained in the use of Engineering Controls and safe work practices.

*Handwashing and handwashing facilities.* Readily accessible handwashing facilities must be provided for occupationally exposed employees. When this is not feasible, antiseptic towelettes or antiseptic hand cleansers used with cloth or paper towels followed by soap-and-water handwashing must be used (e.g., emergency response in remote area of plant). Employees who use gloves or other personal protective equipment must wash their hands immediately (or as soon as possible) after their removal. Similarly, skin must be washed with soap and water and mucus membranes flushed with water immediately after contact with blood or other potentially infectious agents. While not required, antimicrobial soaps (e.g., Dial, Betadine) are recommended.

*Needles and sharps.* Careful handling and disposal of needles and other sharp implements or objects (e.g., scissors, broken glass) cannot be overemphasized. Such practices may represent the single most important protection against bloodborne pathogen transmission. Contaminated needles and sharps shall not be capped, bent, broken, cut, or manipulated (unless, in the case of recapping, specially designed devices for this purpose are used) and must be disposed of in sharps containers as soon as possible after use. Retrieval from sharps containers is prohibited. Containers must be (1) closeable, (2) puncture resistant, (3) leakproof, and (4) color-coded (red or labeled with biohazards symbol).

*Food, cosmetics, contacts.* Eating, drinking, smoking, applying cosmetics or lip balm, handling contact lenses, and the presence of food or beverages are prohibited in areas where Exposure can occur.

*Procedures involving blood or other potentially infectious materials; blood specimen handling and transportation.* All procedures must be performed in a manner that minimizes the likelihood of splashing, splattering, spraying, and forming droplets of blood or other potentially infectious materials (hereinafter "Blood"; e.g., the use of microhematocrit machine with a breakproof observation window). Mouth pipetting is forbidden!

*Engineering Controls checklist*

List Engineering Controls (e.g., handwashing facilities; sharps containers; blood specimen containers; infectious waste containers; handling, storage, and transportation containers) in use in this facility. Complete and/or check all that apply.

[ ] Handwashing facilities
[ ] Sharps containers
[ ] Infectious waste containers
[ ] Handling, storage, and transportation containers
[ ] Regular schedule to examine and maintain or replace Engineering Controls, as needed
[ ] Handwashing facilities readily accessible to occupationally exposed employees (or, when unavailable, antiseptic towelettes or antiseptic hand cleansers with cloth or paper towels followed by soap-and-water handwashing)
[ ] All containers that may be used for storage, transport, or shipping of blood or other potentially infectious materials appropriately labeled or color-coded (biohazard warnings or red bags/containers), leakproof, and themselves not contaminated on the outside
[ ] Review of Engineering Controls at least [____] time(s) per year

Other (list):

[ ] ______________________________
[ ] ______________________________
[ ] ______________________________
[ ] ______________________________
[ ] ______________________________
[ ] ______________________________

*Inspection and maintenance schedule form*

• *Handwashing facilities:* Sinks with soap (preferably antimicrobial) are available at the following locations for use by occupationally exposed employees:

Numerous locations throughout facility
______________________________
______________________________
______________________________
______________________________

Wherever handwashing facilities are not made accessible, antiseptic towelettes [ X ] and/or antiseptic hand cleansers and cloth or paper towels [__] are available for use at the following locations (e.g., in first-aid kits):

Antiseptic towelettes available in first-aid boxes located in rear of the plant and in the office area.

Inspection/maintenance schedule of handwashing facilities: ____[daily] ____

Responsible person/party: ______[name] ____________

• *Sharps containers* (leakproof, puncture-resistant, labeled):

Locations:

At first-aid station

Types of containers used:

See attached.

[ ] Replaced/disposed of when three quarters full?

Responsible person/party: ______[name] ____________

• *Other containers for contaminated materials,* such as glass, garbage pails, laundry (labeled, leakproof, can readily hold contents):

Locations:

First-aid station in office and breakroom

Types of containers used:

Hazardous waste bags

Method of disposal:

______________________________________________________________

Responsible person/party: ______[name] ________________________

[ ] Employees have been informed of and trained in the use of all of the above engineering controls.

Responsible person/party: ______[name] ________________________

*Work Practice Control checklist*

Complete and/or check all that apply.

*Handwashing:*

Employees are informed and trained that they must:

[ ] Wash their hands as soon as possible after removal of gloves and other personal protective equipment (e.g., goggles, face masks).
[ ] Wash their hands with soap and water or flush mucous membranes with water as soon as possible after occupational exposure.

*Sharps:*

[ ] No contaminated needles are bent, recapped, or removed.
[ ] All contaminated, disposable, and reusable sharps (e.g., scissors) are immediately placed in puncture-resistant, labeled or color-coded, leak-proof containers while awaiting disposal or reprocessing/decontamination.
[ ] Employees have been instructed not to overfill containers.

*Personal hygienic practices:*

[ ] Eating, drinking, smoking, applying cosmetics or lip balm, and handling contacts are prohibited in areas where occupational exposure might occur.
[ ] Food and drink are not kept in refrigerators, freezers, shelves, cabinets, or countertops where potentially contaminated materials are or may be present.

*Miscellaneous work practices:*

[ ] All procedures involving blood or other potentially infectious materials are performed in a manner that minimizes splashing, spraying, spattering, and generation of droplets.
[ ] Mouth pipetting is prohibited (if applicable).
[ ] Containers for storage, transport, or shipping of potentially infectious materials are labeled or color-coded and closed prior to leaving the facility (see Hazard Communication).
[ ] Containers with outside contamination are placed within other non-contaminated containers meeting the same required specifications.

[ ] Equipment that may have become contaminated with blood is inspected and decontaminated prior to servicing or shipping (e.g., flexible sigmoidoscope).
[ ] Work Practice Controls are reviewed on a [yearly] basis.
[ ] Employees have been informed and trained in all of the above Work Practice Controls in this checklist.

Responsible person/party: _____[name] ______________________

## *Personal protective equipment*

Personal protective equipment (PPE) must be used when the potential for occupational exposure exists. EMS, Inc., will provide a full complement of PPE.

### *Availability*

Appropriate, properly sized PPE must be provided without cost and made readily accessible to employees when there is the potential for Exposures; for example, gloves, gowns, lab coats, eye protection, masks, faceshields, oropharyngeal airways, resuscitation bags, pocket masks, and overlay barriers are considered appropriate if they protect against infectious materials reaching mucous membranes, skin, and work or street clothes. Hypoallergenic gloves, glove liners, and other alternatives must be made readily accessible to employees who are allergic to standard gloves. It is the company or organization's responsibility to assure that the employee is trained on how to use, and the uses of, the appropriate PPE provided. For examples of recommended PPE by task, see attachment A, this section.

### *Gloves*

Gloves (preferably latex) must be worn when it can be reasonably anticipated that an employee may have hand contact with blood, mucous membranes, and non-intact skin or may handle/touch contaminated surfaces or items. Good hygienic practice also dictates that employees who provide health care and have potentially infectious skin lesions or breaks in their skin protect themselves and their patients by using gloves.

### *Masks; eye and face protection*

Masks in combination with goggles or glasses with solid sideshields or chin-length faceshields must be worn whenever splashes or droplets of blood/infectious materials may be generated and eyes, nose, or mouth contamination might occur.

### *Protective clothing*

Gown, aprons, lab coats, and clinic jackets must be worn in occupational exposure situations according to the task and degree of exposure anticipated.

*Respiratory equipment*

Respiratory equipment such as resuscitation bags, pocket masks with one-way valves, oropharyngeal airways, and/or overlay barriers must be used so that mouth-to-mouth resuscitation is avoided.

*Cleaning, laundering, disposal*

All personal protective equipment must be removed prior to leaving the work area and placed in the appropriate bags or containers for washing, decontamination, or disposal. Should garments be penetrated by blood or other potentially infectious materials, they must be removed immediately and placed in appropriate containers for contaminated laundry. PPE cleaning, repair, and replacement must be performed at no cost to the employee. Home laundering is forbidden. Disposable gloves shall be replaced as soon as possible when contaminated, torn, or punctured and cannot be washed for reuse (utility gloves in good condition may be decontaminated and reused).

*Note:* Training in the proper selection, use, and disposal/decontamination of PPE is of critical importance. Personnel responsible for such PPE training must be qualified and preferably should be experienced in their use. For example, the technique of applying and removing latex gloves to avoid self-contamination should be demonstrated and must be understood and followed by all who may use this form of PPE.

*Replacement personal protective equipment*

Replacement PPE is available at first-aid stations or in the office.

Contact person: ________[name]____________________________

Training responsible party: ________[name]____________________________

### *Attachment A to personal protective equipment*

Recommended PPE by Task

| Task/Activity | Disposable Gloves | Gowns | Masks | Protective Eye Wear |
|---|---|---|---|---|
| Measuring blood pressure, temperature | No | No | No | No |
| Bleeding control (spurting blood) | Yes | Yes | Yes | Yes |
| Bleeding control (minimal bleeding) | Yes | No | No | No |
| Handling and cleaning contaminated instruments | Yes | No, unless soiling is likely | No | No |

*Availability*

The following types of PPE (checked) are available in appropriate sizes for use by employees with potential for occupational exposure:

[ ] Gloves (utility)
[ ] Gloves (latex and/or vinyl)
[ ] Lab/clinic coats
[ ] Gowns
[ ] Eye protection (goggles and/or glasses with sideshields)
[ ] Masks
[ ] Faceshields
[ ] Resuscitation bags
[ ] Pocket masks
[ ] Overlay barriers
[ ] Other (list):
____________________
____________________
____________________
____________________

[ ] PPE is provided without cost and made readily accessible.

*Gloves*

[ ] Gloves are worn when it can be reasonable anticipated that an employee may have hand contact with blood or other potentially infectious materials, mucous membranes, and non-intact skin and during procedures requiring the handling or touching of contaminated surfaces or items.

[ ] Employees who use gloves are trained on when to use them, how to remove them without contaminating themselves, and how to appropriately discard and, in the case of utility gloves, decontaminate them.

Gloves are used during (check all that apply):

[ ] First aid (bleeding)
[ ] Decontamination (spills)

Other:

[ ] ____________________
[ ] ____________________

[ ] Gloves are worn by employees who provide health care to others when the employees have potentially infectious skin lesions or breaks in the skin of their hands.

[ ] Hypoallergenic gloves, glove liners, or other alternatives are made readily accessible to employees who are allergic to standard gloves.

[ ] Disposable gloves are discarded into appropriate, nearby containers as soon as possible after contamination. Disposable gloves are never washed for reuse.

[ ] Utility gloves are available for housekeeping/decontamination.

Gloves appropriate to the task may be obtained by:

[ ] Issuance by or on request of the manager/supervisor

[ ] Use of first-aid/emergency kits which contain them

[ ] Other: ________________________________

If gloves are not used, explain reason why:

______________________________________________________________

______________________________________________________________

______________________________________________________________

______________________________________________________________

*Facial and respiratory protection*

[ ] Masks in combination with goggles or glasses with solid sideshields or chin-length faceshields are used whenever splashes or droplets of blood/infectious materials might contaminate eyes, nose, mouth, or facial skin (e.g., control of spurting blood).

[ ] Resuscitation bags, pocket masks (with one-way valves), and/or overlay barriers are used to avoid mouth and respiratory system contamination during cardiopulmonary resuscitation (CPR).

[ ] Work practices specify under what circumstances, if any, facial protection may be necessary.

Facial/respiratory personal protective equipment may be obtained by:

[ ] Issuance by or on request of the manager/supervisor

[ ] Use of first-aid/emergency kits which contain them

[ ] Other: ________________________________

If facial/respiratory protection is not used, explain reason why:

______________________________________________________________

______________________________________________________________

______________________________________________________________

______________________________________________________________

*Protective clothing*

[ ] Gowns, aprons, and lab coats are available in occupational exposure situations according to task and degree of exposure anticipated (for example, spurting blood, sigmoidoscopic evaluations, decontamination procedures involving large spills of blood).

Protective clothing may be obtained by:

[ ] Issuance by or on request of the manager/supervisor

[ ] Other: ________________________________

[ ] Contaminated protective clothing is placed in appropriate containers as soon as possible after contamination and discarded, decontaminated, or laundered.

Laundering of soiled protective clothing is performed by:

______[company/city/state]________________________________

(*Note:* Employees are not permitted to launder their own contaminated clothing.)

If protective clothing is not utilized, explain reason why:

________________________________________________________

________________________________________________________

________________________________________________________

________________________________________________________

[ ] Employees have been informed and trained in the selection, use, and disposal/decontamination of the above personal protective equipment where applicable.

Responsible person(s)/party(s): _______[name]____________________

Training by: _______[name]____________________

The availability and use of personal protective equipment are reviewed on a _____[yearly]_____ basis by ______[name]________________.

## *Housekeeping*

*Note:* Many locations use outside contractors for their security, housekeeping, laundry, and other services, the performance of which may have the potential for occupational exposures. Therefore, it is advisable to obtain a written agreement that clarifies which organization will be responsible for performing particular tasks having the potential for occupational exposure and under what circumstances and under whose auspices such tasks will be performed. In most instances, it is the company's responsibility to inform outside contractors and/or their employees about the potential for exposure to contaminated materials (e.g., janitorial staff) and it is the outside contractor's responsibility to provide the full protection of the Bloodborne Pathogens standard, including, but not limited to, education, training, and personal protective equipment for their own employees.

### *General*

All worksites must be kept in a clean and sanitary condition. A written schedule for cleaning and a procedure for decontamination of worksites must be available and appropriate for each:

- Location (breakroom, main plant)
- Type of surface to be cleaned (counters, floortile, or carpet)

- Type of soiling present (bloody emesis, small splatter of blood)
- Tasks/procedures performed

*Equipment, environmental, and working surfaces*

All equipment, environmental, and working surfaces must be cleaned and decontaminated with an appropriate disinfectant (e.g., a 1:10 bleach-to-water solution; virucidal or tuberculocidal solutions) as soon as possible after contact with "blood or other potentially infectious materials" ("Blood"), upon completion of procedures that could give rise to contaminated surfaces, and at the end of the work shift if surfaces may have been contaminated since the last cleaning.

*Note:* Disinfectants should be selected based on the Centers for Disease Control Guidelines, MMWR1989, 38 (No. S-6), pages 1–37. Examples include 1 part household bleach to 10 parts water, Steriphene II disinfectant (may be safely used on carpets), Cidex (for cleaning sigmoidoscopic equipment).

Similarly, protective coverings (e.g., plastic wrap, aluminum foil) must be appropriately discarded as soon as possible after visible contamination or at the end of the work shift if contamination could have occurred. All reusable receptacles (bins, pails, pans) must be inspected, decontaminated, and cleaned on a regularly scheduled basis if there is a reasonable likelihood of their being contaminated by Blood and as soon as possible after visible contamination.

*Glassware/reusable sharps*

Broken glassware that may be contaminated must not be picked up directly by hand. A mechanical means of picking up such material (tongs, brush and dust pan, or forceps) must be used in such cases. The mechanical devices themselves must then be decontaminated. Reusable sharps must not be stored or decontaminated in a manner that requires employees to reach by hand into the containers in which they have been placed.

*Regulated waste*

*Contaminated sharps:* Containers for sharps shall be discarded as soon as possible in containers that are (1) closeable, (2) puncture resistant, (3) leakproof (sides and bottom), and (4) properly labeled or color-coded.

Containers for contaminated sharps shall be (1) easily accessible, (2) located as close as possible to the area where they are used or where sharps might be found (e.g., laundries), and (3) kept upright.

When moving sharps containers, they must be (1) closed just prior to moving, and (2) placed in secondary leakproof/puncture-resistant containers (if leakage or puncture is possible), which must also meet all of the required characteristics of the primary container.

Reusable containers must not be opened, emptied, or cleaned in a manner that poses a risk of occupational exposure or percutaneous (through the skin) injury.

*Other contaminated waste:* Other contaminated waste must be handled, contained, and disposed of in the same meticulous fashion as for "contaminated sharps," except for the lack of need for puncture-resistant containers. Routine checks must be performed as necessary for outside contamination and secondary containers used to enclose outside-contaminated primary containers (see Engineering and Work Practice Controls).

*Note:* Disposal of regulated waste must be in accordance with applicable federal, state, and local regulations.

*Laundry:* Contaminated laundry must be (1) handled as little as possible, (2) bagged at the location where it is used (sorting is not permitted), and (3) placed and transported in properly labeled or color-coded containers which prevent soak-through or leakage. Employees who have contact with contaminated laundry must wear protective gloves and other appropriate personal protective equipment (e.g., gowns or aprons). Companies or organizations must use bags meeting the requirements for "other contaminated waste," above, prior to offsite shipping of contaminated laundry.

### *Disposal*

When necessary, EMS, Inc., will contact BFI for disposal of contaminated waste.

### *Attachment A to housekeeping*

*Contaminated spills: recommended procedures*

[ ] Secure area.

[ ] Contact [name] ______________________.

[ ] Determine decontamination equipment necessary:

Personal protective equipment (latex gloves [ ], masks [ ], goggles [ ], gowns [ ], booties [ ]; other ______________________ [ ].

Absorbent materials (e.g., paper towels, commercial absorbent agents) (list):

[ ] ______________________ [ ] ______________________
[ ] ______________________ [ ] ______________________
[ ] ______________________ [ ] ______________________
[ ] ______________________ [ ] ______________________

[ ] 1:10 bleach solution

Other disinfectant solutions (EPA-defined, tuberculocidal) (list):

[ ] ______________________ [ ] ______________________
[ ] ______________________ [ ] ______________________

[ ] Appropriately labeled, puncture-proof/leakproof containers to receive contaminated materials

[ ] Appropriate containers for processing or decontaminating housekeeping/personal protective equipment used

*Decontamination procedure*

[ ] Put on appropriate personal protective equipment.

[ ] Cover spill with absorbent material(s).

[ ] Absorb and containerize grossly contaminated materials.

*Note:* If glass or sharps are present, do not remove directly using hands; use a dust pan and brush and place in an appropriately labeled, puncture-resistant container.

[ ] Disinfect remaining contaminated materials with appropriate chemical agent following manufacturer's instructions.

[ ] Wash/mop area with clean water and/or other appropriate cleansing agent.

[ ] Discard disposable PPE and equipment; containerize all reusable PPE and equipment for appropriate decontamination.

[ ] Wash hands thoroughly.

*Housekeeping checklist*

Schedule:

[ ] A written schedule for cleaning and decontaminating work sites is available.

Housekeeping procedures:

- Cleaning

[ ] All equipment, environmental and working surfaces are cleaned using appropriate disinfectants as soon as possible after known contamination; at end of procedures or work shift if potential for contamination.

[ ] Reusable receptacles are inspected, decontaminated, and cleaned on a regularly scheduled basis and as soon as possible after visible contamination.

[ ] Protective coverings are discarded at the ends of work shifts in which contamination could have occurred and as soon as possible after visible contamination.

- Broken glass

[ ] Broken glass is never picked up directly by hand.

[ ] Mechanical aids (tongs, brush and dust pan, or forceps) are used and appropriately discarded or disinfected before reuse.

- Sharps containers, reusable sharps containers; other contaminated waste

[ ] Reusable sharps containers are never entered by hand.

[ ] Contaminated sharps containers are (1) closeable, puncture resistant, and leakproof (sides and bottom); and (2) properly labeled or color-coded, easily accessible, located as close as possible to the area where they are used or where sharps may be found, and kept upright.

[ ] Sharps containers are not overfilled and are routinely replaced when no more than three quarters full.

[ ] Sharps containers are closed prior to moving.

[ ] Reusable containers are not opened, emptied, or decontaminated in a risky manner.

[ ] Other, non-sharps, contaminated waste is meticulously handled, contained, and disposed of in accordance with applicable federal, state, and local regulations.

- Laundry

Contaminated laundry is

[ ] Handled as little as possible.

[ ] Bagged at the location where it is used (sorting is not permitted).

[ ] Placed and transported in properly labeled or color-coded containers which prevent soak-through or leakage.

[ ] Gloves and, when necessary, gowns or aprons are used when handling contaminated laundry.

[ ] Contaminated laundry is appropriately bagged and labeled (biohazard sign) prior to offsite shipping.

Miscellaneous

- Written/verbal understanding: There is a written or verbal understanding of the potential occupational exposure hazards, decontamination procedures, and responsibilities of the company and its contracted services:

Laundry service (name):

[ ] ______________________________

Other services (name):

[ ] ______________________________

[ ] ______________________________

[ ] ______________________________

Person/party responsible for documenting that the above-mentioned written or verbal understanding has been provided and/or discussed with contracted services (provide name of person):

[ ] ______________________________

[ ] Employees have been informed of, and trained in complying with, all of the above housekeeping requirements.

Person(s)/party(s) responsible:

______________________________________________

______________________________________________

______________________________________________

*Housekeeping schedule*

Worksite(s) (location in facility):

Complete facility cleaned and disinfected daily in accordance with regulations

Type of surface:

Varied surfaces

Methods of decontamination and type of disinfectant (use only approved disinfectants):

See attached list of USDA-approved disinfectants.

Frequency of cleaning/decontamination:

Daily

Responsible person(s)/party:

Manager

## *Hepatitis B vaccine and vaccination series and post-exposure evaluation*

*Note:* Successful compliance with this section must include the completion of all forms. The OSHA Hepatitis B Vaccine Declination Form is mandatory.

### *Hepatitis B vaccination*

The hepatitis B vaccine is a safe, effective immunization against the hepatitis B virus and must be offered free of charge to all employees who may have Exposure after they have participated in a required training program. Employees with Exposure have the option of declining the vaccine; however, any at-risk employee who wishes not to receive the vaccine must sign a Hepatitis B Vaccine Declination Form. If the employer later decides to receive the vaccine, it must still be offered free of charge. The initial vaccine offering must be made no later than July 6, 1992. New employees or employees whose job tasks have changed such that they are now occupationally exposed must be offered the vaccination within 10 working days of beginning the new job task.

Hepatitis B vaccination must be made available to eligible employees at a reasonable time and place and must be given or supervised by a licensed healthcare professional (physician, nurse). EMS, Inc., has designated Dr. Doe at the Family Clinic as its designated physician. The vaccination series currently consists of three separate injections given over a 6-month time period. More than 90% of those vaccinated according to the recommended dosage schedule will develop immunity to the hepatitis B virus. *Note:* The healthcare professional administering the vaccination series must be provided a copy of the OSHA standard.

Employees who have received the HBV vaccination series, are immune to HBV, or have a medical contradiction to receiving the vaccination series are exempt from the requirements of this section. At the present time, booster doses of vaccine are not recommended. If, in the future, a booster dose(s) of hepatitis B vaccine is recommended by the U.S. Public Health Service, such booster dose(s) must be made available free of charge to employees at risk.

### *Post-exposure evaluation and follow-up*

Any employee who has had an "exposure incident" must be offered a post-exposure evaluation and follow-up (including any indicated laboratory tests), free of charge, at a reasonable time and place, under the supervision of a licensed healthcare professional (physician, nurse) and according to current U.S. Public Health Service recommendations. Following an exposure incident, the confidential medical evaluation and follow-up must be made immediately available to the affected employee. Such evaluation and follow-up must include all of the following:

- Documentation of route(s) of exposure and circumstances under which the incident occurred.
- Identification and documentation of the source individual, where feasible (unless prohibited by law).
- Immediate testing of the source individual's blood to determine HBV and HIV status, unless already known (after obtaining written consent; if consent is not obtainable, this must be documented and placed in the employee's medical record). Results of testing must be made available to the affected employee, along with applicable laws/regulations prohibiting disclosure of identity and clinical status of the source individual.
- Determination of affected employee's HBV and HIV status. Blood testing must be initiated as soon as possible after consent is obtained. The employee may elect to have baseline blood drawn but can delay a decision concerning testing on that blood specimen. In such an instance, the blood specimen must be preserved for at least 90 days, whereupon it may be discarded.
- Post-exposure prophylaxis, if indicated, per U.S. Public Health Service guidelines.
- Employee counseling.
- Medical evaluation of any employee-reported illnesses.
- Inclusion of exposure incident on OSHA 200 Injury/Illness Log.

### *Information provided to the healthcare professional*

The company or organization is responsible for providing the healthcare professional who will be performing the employee's hepatitis B vaccination with a copy of the OSHA Bloodborne Pathogens standard. In addition, the healthcare professional evaluating an employee after an exposure incident must be provided:

- Copy of the Bloodborne Pathogens standard
- Copy of this written program
- Description of the employee's duties as they relate to the exposure incident
- Documentation of the routes of exposure and circumstances under which exposure occurred
- Results of the source individual's blood testing, if available
- All medical records relevant to the appropriate treatment of the employee, including vaccination status

### *Healthcare professional's written opinion*

The company or organization must obtain and make available to the exposed employee a copy of the evaluating healthcare professional's written opinion within 15 days of the completion of the evaluation. The written opinion shall be limited to indicating if the employee should receive hepatitis B vaccination and if such vaccination has been given. The healthcare professional's written opinion concerning post-exposure evaluation and follow-up shall be limited to an indication that the exposed employee has been informed of the results of the evaluation and has been told about any medical conditions resulting from Exposure that may require further evaluation or treatment. All of the findings or diagnoses shall remain confidential and not be included in the written report. EMS, Inc., will maintain all medical reports in the individual employee's medical file.

### *Post-exposure response guidelines*

*Decontamination:*

- Skin: wash thoroughly with soap and water.
- Mucous membranes (eyes, nose, mouth): rinse thoroughly with water or normal saline.
- Environmental surface: wash with solution of 1:10–1:100 household bleach and water or other appropriate disinfectant.

*Reporting of exposure incident to designated individual:*

- Document exposure, including route (i.e., needlestick or absorption through mucous membranes [eyes, nose, mouth]).
- Complete OSHA 200 Injury/Illness Log.

*Healthcare professional evaluation:*

- Employee medical evaluation
- Review of exposed employee's medical record relevant to exposure incident (i.e., hepatitis B vaccine status)
- Identification of source individual and, if possible, any HIV and HBV testing, once consent is obtained
- Employee blood testing for HBV and HIV, once consent is obtained (both HBV and HIV prophylaxis needs to be considered in the medical evaluation of any employee who has occupational exposure)

*HIV post-exposure prophylaxis:*

Consider zidovudine (AZT).

*HBV post-exposure prophylaxis:*

| Source Status | Tested Positive | Tested Negative | Not Tested |
|---|---|---|---|
| Unvaccinated | Administer HBIGx1[a] and initiate hepatitis B vaccine series. | Initiate hepatitis B vaccine series. | Initiate hepatitis B vaccine series. |
| Previously vaccinated, known responder | Test exposed person for anti-HBsAG; if adequate,[b] no treatment; if inadequate, give hepatitis B vaccine booster. | No treatment | No treatment |
| Known non-responder | Repeat HBIGx2 or HBIGx1 hepatitis B vaccine. | No treatment | If known high-risk source, may treat as if source were HBsAG positive. |
| Response unknown | Test exposed person for anti-HBsAG; if inadequate, give HBIGx1, plus hepatitis B vaccine series; if adequate, no treatment. | No treatment | Test exposed person for anti-HBsAG; if inadequate, give hepatitis B vaccine booster dose; if adequate, no treatment. |

[a] Hepatitis B immune globulin (HBIG) dose 0.06 mL/kg intramuscularly.

[b] Adequate anti-HBsAG (hepatitis B surface antigen) is greater than or equal to 10 milli-international units.

*Medical counseling and follow-up:*

Written opinion of Dr. Doe will be sent to EMS, Inc., which will forward it to the employee within 15 days of completion of evaluation

*Employees eligible for, and offered, hepatitis B vaccine and vaccination series*

| Employee Name | Date Offered | Accepted? (Y/N) | If Declined, Is Declination Form Signed? (Y/N) |
|---|---|---|---|
| ______________________ | ________ | ____ | ____ |
| ______________________ | ________ | ____ | ____ |
| ______________________ | ________ | ____ | ____ |
| ______________________ | ________ | ____ | ____ |
| ______________________ | ________ | ____ | ____ |
| ______________________ | ________ | ____ | ____ |
| ______________________ | ________ | ____ | ____ |
| ______________________ | ________ | ____ | ____ |
| ______________________ | ________ | ____ | ____ |
| ______________________ | ________ | ____ | ____ |
| ______________________ | ________ | ____ | ____ |
| ______________________ | ________ | ____ | ____ |

*Hepatitis B Vaccine Declination Form ECP VIII-2 (mandatory)*

I understand that due to my occupational exposure to blood or other potentially infectious materials I may be at risk of acquiring hepatitis B virus (HBV) infection. I have been given the opportunity to be vaccinated with the hepatitis B vaccine, at no charge to myself; however, I decline hepatitis B vaccination at this time. I understand that by declining this vaccine, I continue to be at risk of acquiring hepatitis B, a serious disease. If, in the future, I continue to have occupational exposure to blood or other potentially infectious materials and I want to be vaccinated with the hepatitis B vaccine, I can receive the vaccination series at no charge to me.

Name (please print):______________________________________

Title:______________________________________

Signature: ______________________________________

Date: ______________________________________

Site/facility location: ______________________________________

Place in confidential medical record.

*Employee exposure incident checklist*

*Note:* This checklist is to be used when an employee sustains an "exposure incident" (see Definitions section, above).

*Employee actions:*

[ ] Decontamination of skin, mucous membranes or environmental surface

[ ] Notification of exposure incident to designated individual

*Responsible person(s)/party(s) actions:*

[ ] Documentation of exposure incident by completing Employee Exposure Incident Record

[ ] Completion of OSHA 200 Injury/Illness Log

[ ] Identification and documentation of source individual where feasible, unless prohibited by law

[ ] HBV and HIV testing of source individual's blood after obtaining written consent (unless already known); document if source individual's consent is not obtained

[ ] Results of source individual's test results made available to affected employee with prohibitions against disclosure of source individual's identity or clinical status

[ ] Determination of affected employee's HBV and HIV status (complete Employee Exposure Incident Record)

*Information provided to the health professional evaluating an affected employee after exposure incident:*

[ ] Copy of Bloodborne Pathogens standard

[ ] Employees duties as they relate to exposure incident

[ ] Routes of exposure and circumstances under which exposure occurred

[ ] Results of source individual's blood testing, if available

[ ] All relevant medical records

[ ] Medical evaluation and treatment as outlined in the post-exposure response guidelines section, above.

[ ] Written medical opinion sent to company or organization management and forwarded to affected employee within 15 days of completion

## *Employee exposure incident record*

Date and time of exposure incident:

______________________________________________

Name of exposed employee:

______________________________________________

Exposure incident reported to:

______________________________________________

Route(s) of exposure:

___

Circumstances of exposure incident (use additional sheet, if necessary):

___

___

___

___

Medical evaluation performed by:

___

Employee vaccination status:

___

Employee HBV/HIV status:

___

Medical evaluation and treatment (including post-exposure prophylaxis and counseling):

___

Written opinion of medical evaluation sent to company or organization management and forwarded to exposed employee on (date):

___

Place in confidential medical record.

## *Hazard communications*

Warning labels and signs must be used to identify items that can pose a hazard. These labels and signs will ensure that anyone who may come in contact with objects so identified know that they must handle the objects with care. OSHA has chosen the fluorescent biohazard sign as the appropriate symbol and red as the designated hazard color code. Labels may be attached by string, wire, adhesive, or any other method that will not allow them to fall off accidentally. They can also be a part of a container itself. Red bags or red containers may be substituted for labels.

### *Biohazard sign*

Containers of regulated waste, refrigerators or freezers that hold potentially infectious materials (e.g., blood), and other containers used to transport or store blood or infectious materials must be labeled with the biohazard sign or be color-coded (red bags or red containers). Contaminated equipment must be labeled so as to designate which portions, if any, remain contaminated. Individual containers that are placed inside another labeled container for storage or transport do not need to be labeled separately. Regulated waste that has been decontaminated does not need any labeling or color-coding.

### *Hazard communications checklist*

[ ] Warning labels are affixed to containers of regulated biomedical waste, refrigerators, and freezers containing blood or other potentially infectious material. Containers used to store, transport, or ship blood or other potentially infectious materials are labeled in a similar fashion.

[ ] Warning labels use the OSHA-designated biohazard sign that is fluorescent orange or orange-red with letters or symbols in a contrasting color.

[ ] Warning labels are either a part of the container itself or attached by string, wire, adhesive, or any other method that will not allow them to fall off accidentally.

[ ] Red bags or red containers are substituted, where appropriate, for labels. The hazards of blood and other potentially infectious material have been communicated to all employees who have occupational exposure in this facility and the use of appropriate warning labels and signs has been explained.

Responsible person(s)/party: ______[name]__________________________

## *Training and education*

The standard has mandated that EMS, Inc., provide all their employees determined to have occupational exposure an educational training program, free of charge and conducted during normally scheduled working hours. It may be helpful when conducting these training sessions to prepare examples of signs and labels that are required (e.g., biohazard labels), have available articles of personal protective equipment to demonstrate their proper use (e.g., gloves, gowns, face masks, mouthpieces for CPR), and demonstrate the work practices that are essential to this program (e.g., disposal of sharps into sharps container). Training videos are available. These may be used as an aid in the training session but are not to be used in place of an individual trainer. The training program must be provided to an employee at the time of initial employment or before June 4, 1999, and must be provided at least annually thereafter to all occupationally exposed employees. Additional training must be provided when new or modified tasks or procedures may affect an employee's occupational exposure. These training records must be

- Kept for three (3) years from the date of the training session.
- Kept available upon request to all employees or their representatives.
- Made available upon request to OSHA.

Examples of training program videos include:

*Bloodborne Pathogens*
Summit Training Service, Inc.
(616)784-4500; (800)842-0466
Time: 12 minutes
Excellent for general workplace settings

*Preventing Bloodborne Disease*
Tel-A-Train
(615)266-0113
Time: 18 minutes
Excellent, especially for first responders

*Ems, Inc., training program requirements*

The training program provides the employee with the following information:

1. Accessible copy of the regulatory text of this standard and an explanation of its contents
2. General explanation of the epidemiology and symptoms of bloodborne diseases
3. Explanation of the modes of transmission of bloodborne pathogens
4. Explanation of the Exposure Control Plan and where the employee can obtain a copy of the written plan
5. Explanation of recognizing tasks and other activities that may involve exposure to potentially infectious materials
6. Explanation of the use and limitations of methods to prevent or reduce exposure, including appropriate Engineering Controls, Work Practice Controls, and personal protective equipment
7. Information on the types, proper use, location, removal, handling, decontamination, and disposal of personal protective equipment
8. Explanation of the basis for selection of personal protective equipment
9. Information on the hepatitis B vaccine, including information on its efficacy, safety, and method of administration; benefits of being vaccinated; and the vaccine and vaccination being offered free of charge
10. Information on the appropriate actions to take and persons to contact in an emergency involving blood or other potentially infectious materials
11. Explanation of the procedure to follow if an exposure incident occurs, including the method of reporting the incident and the medical follow-up that will be made available
12. Information on the post-exposure evaluation and follow-up that the employer is required to provide for the employee after an exposure incident
13. Explanation of the signs and labels and/or color-coding required to identify potentially infectious materials
14. Opportunity for interactive questions and answers with the person conducting the training session

*Bloodborne pathogens training program*

**Bloodborne Pathogens Training Record**

Facility location: Date:

______________________________ ______________________________

Trainer's name: ______________________________

Trainer's qualifications: ______________________________

______________________________

Content (check all that apply):

[ ] Items 1–14 listed in training program requirements section, above

[ ] Q and A with trainer

[ ] Post-training test evaluation

Instructional materials (check all that apply):

[ ] Training program document

[ ] Bloodborne Pathogens standard

[ ] Videocassette

[ ] Other:________________________________________

________________________________________________

________________________________________________

Trainees (use additional pages, as needed):

| Name | Job Classification/Title |
|---|---|
| See attached. | |
| | |
| | |
| | |
| | |
| | |
| | |
| | |
| | |
| | |
| | |
| | |
| | |
| | |
| | |
| | |
| | |
| | |

**Bloodborne Pathogens Training Program**

*A. Bloodborne Pathogens standard*

Copies of the *Bloodborne Pathogens Manual*, this training program, and the OSHA standard upon which they are based are located in the office and are readily available for use by all employees. Contact [name].

*B. Epidemiology and symptoms of bloodborne diseases*

Disease-producing organisms (pathogens) may be found in the blood and certain other body fluids. The pathogens can be transmitted by various means, such as breaks in the skin caused by a needlestick or any sharp instrument, by absorption through skin that has been abraded, or absorption through mucous membranes of the eyes, nose, or mouth. Two disease-producing organisms that are of most concern are the hepatitis B virus (HBV) and the human immunodeficiency virus (HIV). Although the focus of OSHA's standard is directed toward HBV and HIV, it is important that you realize there are other potentially infectious organisms present in blood; following the instructions in this training program for protection against HBV and HIV will also provide you with risk reduction against all bloodborne pathogens.

- Hepatitis B

The acute and chronic consequences of hepatitis B virus (HBV) infection are major health problems in the U.S. An estimated 200–300,000 new infections occurred annually during the period 1980–1991. Approximately 8700 healthcare workers each year contract HBV, and about 200 will die as a result. Transmission of HBV is by direct injection through the skin (e.g., needlestick injuries in healthcare workers; the use of shared needles and syringes by intravenous drug users), blood contamination of mucous membranes (e.g., eyes, mouth, nose), exchange of sexual fluids during intercourse (e.g., semen and vaginal secretions), and transmission from an infected mother to her newborn infant. Infection with HBV may result in a variety of clinical symptoms from asymptomatic infection to a mild, flu-like illness to a very severe infection that may result in debilitating hepatitis, cirrhosis of the liver, or primary liver cancer. An individual who becomes infected with HBV may develop a chronic form of hepatitis from which they never completely recover. Currently, an estimated 1.25 million persons in the U.S. have chronic HBV infection. Persons infected with HBV are capable of transmitting this organism during any of the clinical states already mentioned. Once an individual is infected with HBV, the presence of certain "markers" can be determined (HBsAG, hepatitis B surface antigen; HBeAG, hepatitis B envelope antigen; HBsAb, hepatitis B surface antibody). HBsAG and HBeAG may give us information that an individual is still in the infectious stage of disease and capable of transmitting the hepatitis virus. HBsAb may show that the infected individual has produced antibodies against the hepatitis virus, which indicates recovery from infection. These antibodies are "protective antibodies" and prevent the individual from becoming ill if exposed to HBV again. These protective antibodies are what is produced in individuals who receive the hepatitis B vaccination series. The risk of death and impairment of health resulting from acute and chronic hepatitis B infection is significant. Therefore, knowledge of the disease and prevention against it are important. The best defense against HBV is to receive the hepatitis B vaccination series.

- Human immunodeficiency virus

The organism that causes the Acquired Immune Deficiency syndrome (AIDS) is the human immunodeficiency virus (HIV). This virus was isolated during 1983–1984. HIV transmission can occur by the same methods as described for HBV transmission. Symptoms of HIV infection vary according to disease progression. After infection with HIV, the individual may experience an illness that is characterized by fever, enlarged lymph nodes, muscle and joint pains, diarrhea, fatigue, and possibly a rash. This acute illness is usually self-limiting and resolves within 2 to 4 weeks. The majority of individuals with HIV develop antibodies that are antigenic markers that can be found in the blood after infection with HIV. These are not "protective antibodies" as previously discussed for hepatitis B virus infection. Most persons who are infected with HIV may remain without symptoms for months to years following infection; however, the majority will eventually develop AIDS. HIV infection affects the immune system such that the infected individual becomes susceptible to a wide range of illnesses. These may include infections from bacterial, fungal, and parasitic organisms that an individual with a normal immune system would only rarely experience. AIDS patients may also be susceptible to developing various cancers and quite often experience neurologic complications. Any of these clinical disorders can be aggressive, rapidly progressive, difficult to treat, and less responsive to traditional methods of therapy. Transmission of HIV from an infected individual may occur throughout any stage of the disease. There is no vaccine against the HIV; therefore, the best protection is close adherence to the specific measures outlined in this training program to reduce the risk of infection from all bloodborne pathogens.

*C. Modes of transmission of bloodborne pathogens*

Bloodborne infections may enter the body in the work environment through a break in the skin caused by a needlestick or any sharp instrument that has been contaminated with potentially infectious materials; by absorption through skin that is inflamed, abraded, or cut; or by absorption through mucous membranes in the mouth, nose, or eyes. Nonoccupational transmission of bloodborne infections can also occur through high-risk behavior, such as intravenous drug abuse and certain sexual activities.

*D. Exposure Control Plan*

The Exposure Control Plan has been designed to protect healthcare workers against potentially infectious materials. The objective of the plan is to provide employees with the education and training necessary to reduce and/or eliminate exposure to bloodborne pathogens. The Exposure Control Plan consists of several parts, as follows.

- Exposure determination

Certain employees, by virtue of their job classification and/or title, have occupational exposure to blood and other potentially infectious materials.

These include physicians, nurses, laboratory technicians, or any other employees with a similar risk to exposure. Certain employees may not normally be exposed to potentially infectious materials, but as part of their job duties may occasionally have contact with blood or other potentially infectious materials. These include CPR-trained employees (e.g., supervisors) who, as part of their job duties, act as first-responders to emergencies in the work environment; housekeeping staff who clean medical clinics or other areas where occupational exposure may occur; or any other employee who may occasionally have contact with potentially infectious materials.

- Methods of compliance

Universal Precautions must be observed to prevent contact with blood or other potentially infectious materials. Under circumstances in which differentiation between body fluid types is difficult or impossible, all body fluids shall be considered potentially infectious materials.

- Engineering Controls and Work Practice Controls

Engineering and Work Practice Controls are used to eliminate or minimize employee exposure. Where occupational exposure remains after institution of these controls, personal protective equipment should also be used.

- Personal protective equipment (PPE)

When there is potential for occupational exposure, the company will provide, at no cost to the employee, appropriate personal protective equipment such as, but not limited to, gloves, gowns, laboratory coats, faceshields or masks and eye protection, mouthpieces, resuscitation bags, pocket masks, or other ventilation devices.

- Housekeeping

All worksites are to be maintained in a clean and sanitary condition. Each worksite will be cleaned on a specified schedule. General procedures for cleaning and disinfecting minor and large spills are outlined in section J.

- Medical waste disposal

The procedures for disposal of infectious waste will be designed to protect the health and safety of the healthcare employee while adhering to federal, state, and local regulations.

- Hepatitis B vaccination and post-exposure evaluation and follow-up

Hepatitis B vaccination series will be offered free of charge to those employees who have been determined to have occupational exposure to blood and other potentially infectious materials. This vaccine series, along with the safe work practices covered in this training program, is one of the most effective ways to prevent transmission of HBV. In addition, a plan is in place should you have an exposure incident. This plan includes a medical evaluation, specific treatment when necessary, and medical follow-up.

*E. Tasks/activities with occupational exposure*

The tasks and activities that may involve exposure to blood or other potentially infectious materials in a given job classification or job title are used to make the exposure determination of those employees at potential risk.

*F. Methods of compliance*

Methods of compliance are various practices and/or procedures that are to be followed to minimize the potential for occupational exposure. Methods of compliance include: (1) Universal Precautions; (2) Engineering and Work Practice Controls; (3) personal protective equipment; (4) housekeeping; and (5) laundry.

- Universal Precautions

Universal Precautions is a system of infection control techniques developed by the Centers for Disease Control (CDC) designed to protect employees from exposures to blood and other potentially infectious materials. Because bloodborne infections can be transmitted from apparently healthy persons who have no outward signs or symptoms of a disease, you should therefore assume that *body fluids from all patients are potentially infectious.*

- Engineering Controls and Work Practice Controls

Engineering Controls are methods used to eliminate or minimize occupational exposure. These can include self-sheathing needles, conveniently placed sharps containers and red bag containers for disposal of potentially infectious materials, handwashing facilities, and appropriately labeled (e.g., biohazard warning) cabinets or storage containers for blood or other potentially infectious materials. Eating, drinking, smoking, applying cosmetics or lip balm, and handling contact lenses are prohibited in work areas where there is the likelihood of occupational exposure. Food and drink must not be kept in refrigerators, freezers, shelves, cabinets, or on countertops or benchtops where blood or other potentially infectious materials are present. Work Practice Controls reduce the likelihood of occupational exposure by assuring that procedures are properly performed:

*Handwashing facilities:* Employees must wash their hands immediately, or as soon as possible, after removal of gloves or other PPE. Gloves should not be considered as a substitute for handwashing. If handwashing facilities are not readily available (e.g., when responding to the scene of an emergency), an antiseptic hand cleanser with a cleaning cloth or paper towels or antiseptic towelettes will be provided in the emergency equipment and must be used. When antiseptic cleansers or towelettes are used, hands must be washed with soap and water as soon as possible.

*Management of sharps:* Sharps containers must be puncture resistant, labeled with the universal biohazards symbol or color-coded in red, leakproof on the sides and bottoms, and translucent, to determine the amount of material in the container. Contaminated needles and sharps are not to be recapped or removed unless no alternative is available. If recapping or needle removal is to be done, it must be done with the use of a mechanical device or a one-handed technique.

*Personal protective equipment (PPE):* Where there is potential for occupational exposure, the company will provide at no cost to the employee appropriate PPE such as, but not limited to, gloves, gowns, laboratory coats, faceshields or masks and eye protection, mouthpieces, resuscitation bags, pocket masks, or other ventilation devices. When PPE use is required, it will be provided in appropriate sizes and will be readily available. For special needs (e.g., requiring the use of hypoallergenic gloves or powderless gloves), accommodations will be made and the equipment will be available.

EMS, Inc., will clean, launder, and dispose of PPE provided to employees, free of charge. *Employees are not allowed to launder their own PPE.* Never leave a work area without first removing any PPE and disposing of it in the appropriate containers.

Disposable gloves must be worn when it can be reasonably anticipated that you may have skin or mucous membrane contact with blood or other potentially infectious materials (e.g., when drawing blood, handling vaginal speculums, cleaning sigmoidoscope). As soon as a task is completed, gloves are to be removed and discarded in the appropriate containers, and hands must be washed. Disposable (single-use) gloves are not to be washed or decontaminated for reuse. For general housekeeping activities, utility gloves may be decontaminated for reuse; however, utility gloves must be discarded if they are cracked, peeled, torn, punctured, or in a state of deterioration.

Masks in combination with shields such as goggles or glasses with solid side panels or chin-length face shields are worn whenever splashes, sprays, spatters or droplets of blood, or other potentially infectious materials may be anticipated and eye, nose, or mouth contamination could occur. Prescription eyeglasses with solid sideshields are appropriate.

Appropriate protective clothing may be gowns, aprons, lab coats, clinic jackets, or similar clothing. The choice of what clothing to wear will depend upon the task and risk of exposure anticipated (e.g., gloves only for blood drawing).

Mechanical respiratory assist devices such as mouthpieces, bag valve masks, pocket masks, and other such CPR aids will be available to all personnel who have trained in CPR for resuscitation use.

*Housekeeping:* All worksites are to be maintained in a clean and sanitary condition according to a schedule based upon location, type of surface, type of contamination, and task or procedures being performed. All equipment and surfaces must be cleaned and decontaminated after contact with blood or other potentially infectious materials. The use of disinfectants is based upon the Centers for Disease Control (CDC) guidelines and includes common household bleach and Cidex. Protective coverings such as plastic wrap, aluminum foil, or other imperviously backed absorbent paper may be used to cover equipment or environmental surfaces. Coverings are to be removed and replaced as soon as feasible (e.g., between patients) when they have become contaminated or at the end of the work shift, if they have become contaminated during the shift. PPE must be worn for all decontamination procedures. Employees are strictly prohibited from placing hands into receptacles to retrieve any materials. Broken glass or other sharp items should be removed using mechanical devices, such as a brush and dustpan or forceps. Mechanical devices are either disposed of or decontaminated after use. All receptacles (reusable) that have a reasonable likelihood for becoming contaminated with blood or other potentially infectious materials are inspected on a regular basis and decontaminated when appropriate.

*G. Personal protective equipment*

See personal protective equipment section, above.

*H. Basis for selection of personal protective equipment*

1. *Gloves:* In performing tasks where you can reasonably anticipate hand contact with blood or other potentially infectious materials, you must wear gloves.
2. *Face protection:* Face protection is required when droplets of blood or other potentially infectious materials may splash or spray during a procedure. When indicated, appropriate covering of the face is important to ensure that the eyes, nose, and mouth are shielded.
3. *PPE:* This includes gowns, aprons, lab coats, clinic jackets, and other similar garments whose use by personnel would depend upon the task and degree of exposure anticipated.
4. *Respiratory equipment:* Mouthpieces, resuscitation bags, and other ventilatory equipment must be made available so that mouth-to-mouth resuscitation is avoided. Reusable mouthpieces must be cleaned and disinfected prior to reuse.

A list of recommended uses of PPE is found on the next page.

*I. Hepatitis B vaccine*

The hepatitis B vaccine is a safe, effective measure available to protect healthcare workers from hepatitis B infection. This vaccination series is offered free of charge. You are not required to receive the vaccine; however, if you choose not to accept this offer you must sign a form declining the vaccination series. Even though you may decide to decline the offer at this time, you still will be able to request the vaccination series at any future date. The hepatitis B vaccine is a noninfectious vaccine prepared genetically from yeast cultures rather than human blood or plasma; therefore, there is no risk of contamination from other bloodborne pathogens nor is there any chance of developing hepatitis B from the vaccine. The vaccination series consists of three separate infections given over a 6-month time period. The second injection

Examples of Recommended Personal Protective Equipment for Worker Protection Against HIV and HBV Transmission

| Task/Activity | Disposable Gloves | Gowns | Masks | Protective Eye Wear |
|---|---|---|---|---|
| Measuring blood pressure, temperature | No | No | No | No |
| Bleeding control (spurting blood) | Yes | Yes | Yes | Yes |
| Bleeding control (minimal bleeding) | Yes | No | No | No |
| Handling and cleaning contaminated instruments | Yes | No, unless soiling is likely | No | No |
| Emergency childbirth | Yes | Yes | Yes, if splashing is likely | Yes, if splashing is likely |
| Endotracheal intubation, esophageal obturator use | Yes | No | No, unless splashing is likely | No, unless splashing is likely |
| Oral/nasal suctioning, manually cleaning airway | Yes | No | No, unless splashing is likely | No, unless splashing is likely |

Reprinted from Department of Health and Human Services (NIOSH) Centers for Disease Control, *Guidelines for Prevention of Transmission of HIV and HBV to Health Care and Public Safety Workers,* HHS Publications No. 89-107, 1987, Table 4, p. 28.

is given one month after the first one, and the third injection, which completes the vaccination series, is given 6 months after the first. It is important for individuals to receive all three injections, as more than 90% of those vaccinated according to this recommended dosage schedule will develop immunity to the hepatitis B virus.

*J. Appropriate actions to take and persons to contact in an emergency involving blood or other potentially infectious materials*

When an exposure incident occurs, the area is to be cleaned and decontaminated as soon as possible.

- Cleaning and decontaminating a blood spill

The appropriate use of PPE must be followed, and, of course, gloves will always be essential. First, wipe the spill with absorbent paper towels and dispose of them in the appropriate container. Wipe the area with a disinfectant approved by the USDA and EPA (antibacteriocidal, antituberculocidal) or a solution of one part bleach to ten parts water. Dispose of paper towels, remove any PPE, and wash hands thoroughly with soap and water.

- Decontaminating laboratory/treatment room

Standard cleaning and decontamination procedures apply to patient care areas. You must always wear gloves and use EPA- and USDA-approved disinfectants. If disposable coverings are used on surfaces and objects in the room (e.g., exam tables, work carts), they must be replaced immediately after becoming contaminated. These coverings should be discarded and replaced with clean coverings after use with each patient.

*K. Occupational exposure/post-exposure evaluation and follow-up*

After a needlestick, splash, or other body fluid exposure, the exposed area must be washed immediately. For skin contact, the area should be washed thoroughly with soap and water. For mucous membrane contact, flush the area thoroughly with water or normal saline. You must notify the individual who has been assigned the responsibility for filing these incidents. If this person is not a healthcare professional (e.g., physician, nurse), then the employee must be directed to the designated healthcare professional who will perform the post-exposure confidential evaluation and follow-up.

*L. Information required by employer to provide to employee filing exposure incident*

If an exposure incident occurs, you will be provided with:

1. Blood testing for HIV and HBV as soon as possible. If you choose to delay testing, the initial specimen will be held for 90 days, within which you may elect to have HIV testing done; otherwise, the sample is discarded after 90 days. Refusal to have HIV and HBV testing must be documented in your record.
2. Post-exposure treatment, when medically indicated.
3. Medical follow-up, including evaluation of illnesses (especially those associated with fevers) that may occur primarily within 6 to 12 weeks after the exposure event.
4. The evaluating healthcare professional's written opinion within 15 days of completion of the evaluation. This written opinion will state that: (a) you have been informed of the results of the evaluation, and (b) you have been informed about any medical conditions resulting from exposure to blood or other potentially infectious materials which require further evaluation or treatment. All other findings or diagnoses will be discussed only with you and kept confidential in your medical record and are not to be included in the written report (e.g., diagnosis of a positive HIV or HBV blood test).

*M. Signs, labels and/or color-coding to identify potentially infectious materials*

Warning labels must be placed on all containers with infectious waste. These include refrigerators and freezers containing blood or other potentially infectious materials and any containers used to store, transport, or ship blood or other potentially infectious materials. Labels displaying the universal biohazard precaution symbol are required. These labels must be fluorescent orange or orange-red with letters or symbols in a contrasting color. They must be put on the containers by any method (e.g., adhesive, wire) that prevents their loss or unintentional removal. For all non-sharp contaminated materials, a red bag may replace the use of the biohazard symbol.

*N. Question-and-answer period*

The person(s) conducting the training session will provide ample opportunity for each trainee to ask questions that may have been generated during the training program. A sample post-training quiz is provided below.

**Bloodborne Pathogens Post-Training Evaluation Quiz**

1. Pathogens (disease-producing organisms) can be transmitted by:
   a. Blood.
   b. Absorption through membranes of the mouth.
   c. Absorption through membranes of the eyes.
   d. All of the above.
2. Hepatitis B virus (HBV) is:
   a. Not a problem for healthcare workers.
   b. Not transmitted efficiently by infected blood.
   c. The virus that causes AIDS.
   d. Highly preventable by hepatitis B vaccination.
3. Human immunodeficiency virus (HIV):
   a. Is highly preventable by hepatitis B vaccination.
   b. Is only transmitted by blood.
   c. Affects the immune system.
   d. Affects only intravenous drug users and either homosexual or bisexual men.
4. Hepatitis B vaccination:
   a. Must be taken by all healthcare employees.
   b. Consists of one injection.
   c. Is genetically prepared from yeast and carries no risk of transmitting hepatitis B virus.
   d. Has a minimal charge for administration to all healthcare employees.
5. Personal protective equipment (PPE):
   a. May be gowns, gloves, and masks.
   b. Is provided by the employer at no cost to the employee.
   c. Must be conveniently located and available.
   d. All the above.
6. Disposal of infectious waste:
   a. Can be done in any container, as long as it closes tightly.
   b. Can be done along with regular noninfectious waste.
   c. Must be labeled with a biohazard label or contained in a red bag.
   d. Must occur at the end of each work shift no matter how much or how little waste there is.

7. Handwashing facilities:
   a. Will always be readily available.
   b. Do not have to be used as long as gloves are worn.
   c. May consist of antiseptic cleansers or towelettes until soap and water are available.
   d. Are only necessary when blood drawing occurs.
8. An exposure incident:
   a. May require medical treatment.
   b. Is reported on the OSHA 200 Injury/Illness Log.
   c. Is evaluated by a healthcare professional.
   d. Must be documented in the employee's medical record.
   e. All of the above.
9. The best way to prevent HBV infection is:
   a. Always wear gloves when touching patients.
   b. Receive the HBV vaccination.
   c. Use a mouthpiece during CPR resuscitation.
   d. Receive the HIV vaccination.
10. If a spill of blood occurs:
    a. Leave the area immediately and call housekeeping to clean it up.
    b. Immediately put on the appropriate PPE and wipe up the spill, disinfect the area, and dispose of all infectious waste appropriately.
    c. Immediately clean up using a 1:10 bleach-to-water solution.
    d. Wait until the spill dries before attempting any clean up.
11. If recapping contaminated needles is necessary, you should always use the one-handed scoop method or a recapping device to prevent needlestick injury.
    a. True.
    b. False.
12. Universal Precautions should be observed when working with which group?
    a. Male homosexuals.
    b. Only patients with AIDS.
    c. Drug users.
    d. All patients.
13. Masks and protective eyewear are designated to protect you from:
    a. Needlestick injury.
    b. Clothing contamination.
    c. Mucous membrane contact.
    d. All of the above.
14. Clearly marked, puncture-resistant containers should be available to dispose of used needles or other disposal sharps.
    a. True.
    b. False.

15. Which activity can spread HIV or HBV from one person to another outside of work?
    a. Using a toilet.
    b. Giving blood.
    c. Shaking hands.
    d. Having sex.

*Answer key*
1. d; 2. d; 3. c; 4. c; 5. d; 6. c; 7. c; 8. d; 9. b; 10. b; 11. a; 12. d; 13. d; 14. a; 15. d.

## *Recordkeeping*

### *Medical records*

Accurate medical records for each employee with Exposure must be maintained. These medical records should include:

- Name and social security number of the employee
- Employee's hepatitis B vaccination status, including the dates of all the hepatitis B vaccinations and any medical records relative to the employee's ability to receive the vaccination (e.g., the employee has previously received the complete hepatitis B vaccination series or the vaccine is contraindicated for medical reasons)
- Signed Declination Form (ECP VIII-2) if the employee has declined the hepatitis B vaccination offer

If an exposure incident should occur, the records should include:

- Post-exposure record, which includes circumstances of the exposure incident, the medical evaluation, testing and treatment, results of examinations, and follow-up
- Healthcare professional's written opinion
- Information provided to the healthcare professional

The employee's medical record must be kept confidential and not disclosed or reported without the employee's written consent to any person within or outside the workplace except as required by this section (e.g., to the evaluating healthcare professional) or as may be required by law. These medical records must be kept for the duration of employment plus 30 years.

### *Training records*

Training records must include the following information:

- Dates of the training sessions
- Contents or summary of the training session and/or program
- Names and qualifications of persons conducting the training sessions
- Names and job titles of all persons attending the training sessions

The training records must be maintained for 3 years from the date on which the training session occurred.

*Availability*

The employee medical and training records must be provided upon request for examination and copying to the subject employee, to anyone having the written consent of the subject employee, to OSHA, and to the National Institute for Occupational Safety and Health in accordance with the provisions in OSHA standard 29 C.F.R. 1910.20.

*Transfer of records*

If a company ceases to do business and there is no successor employer to receive and retain the records, the company or organization must make appropriate arrangements for their retention. The company or organization must determine if the records can be retained within another facility. The Director of NIOSH must be informed of the arrangements that have been made for retention of records.

*Recordkeeping checklist*

[ ] Medical records are maintained for each employee with occupational exposure.

Medical records include:

[ ] Name and social security number of the employee

[ ] Employee's hepatitis B vaccination status, including the dates of all HBV vaccinations and any medical reason for the vaccine being contraindicated

[ ] Signed declination of HBV vaccination offer, if the employee did so

[ ] All results of examinations, medical testing and follow-up procedures

[ ] Healthcare professional's written opinion/evaluation

[ ] Medical records are kept confidential and not disclosed or reported without the employee's express written consent.

[ ] Medical records are kept for the duration of employment plus 30 years.

Training records include:

[ ] Dates of training sessions

[ ] Contents/summary of training session

[ ] Name(s)/qualifications of trainer(s)

[ ] Names and job titles/classification of all persons attending training session

[ ] Training records are maintained for 3 years from the date on which training occurred.

Responsible person(s)/party: ______[name]______________________

## Bloodborne pathogens program

This OSHA standard was developed to protect employees from possible exposure to HIV, hepatitis, and other bloodborne diseases that the employee may be exposed to in the workplace. Additionally, this standard requires the proper labeling and disposal of all medical waste, which includes a wide range of items from a blood sample to a used bandage. Below is the recommended procedure for the development of a program to achieve compliance with this standard.

1. Acquire a copy of the OSHA standard.
2. Review this standard with your management team and acquire management commitment and appropriate funding for the program.
3. Develop a written program incorporating all required elements of the standard which include, but are not limited to, Universal Precautions, Engineering and Work Practice Controls, personal protective equipment, housekeeping, infectious waste disposal, laundry procedures, training requirements, hepatitis B vaccinations, information to be provided to the physician, medical recordkeeping, signs and labels, and availability of medical records.
4. In regard to Universal Precautions, your operations should be analyzed to provide all necessary safeguards to employees who may have possible contact with human blood. Engineering Controls and Work Practice Controls should be examined and evaluated. Procedures should be established for the safe disposal of used needles, used personal protective equipment, and other equipment. Such areas as the refrigerator, cabinets, or freezers where blood could be stored should be prohibited for the storage of food or drink (e.g., employee's keeping their lunches in the nurse's refrigerator). Employees should be properly trained in the requirements of this standard. Employees must be prohibited from eating, drinking, smoking, applying cosmetics or lip balm, or handling contact lenses after possible exposure.
5. Where there is the potential for exposure, personal protective equipment such as surgical gloves, gowns, fluid-proof aprons, faceshields, pocket masks, ventilation devices, etc. must be provided to employees. This requirement is especially important for first-aid responders and plant medical personnel who may be exposed when providing first-aid to an injured worker. The personal protective equipment must be appropriately located for easy accessibility. Hypoallergenic personal protective equipment should be made available to employees who may be allergic to the normal personal protective equipment.
6. Employers are required to maintain a clean and sanitary work place. A *written* schedule for cleaning and sanitizing all applicable work areas must be implemented and included in your program. All areas, equipment, etc. that have been exposed (e.g., after an accident) must

be cleaned and disinfected. Exposed broken glass must *not* be picked up directly with the hands. A dust broom, vacuum, tongs, or other instrument must be used.

7. All infectious waste, such as bandages, towels, or other items exposed to human blood, must be placed in a closeable, leakproof container or bag with the appropriate label. This waste must be properly disposed of as medical waste according to federal, state, and local regulations. Check with local hospitals or governmental agencies to acquire the regulations applicable to your facility.
8. Contaminated uniforms, smocks, and other items of personal clothing must be laundered in accordance with the standard. *Do not send contaminated uniforms, smocks, etc. to the in-plant laundry.*
9. Appropriate employees should be trained in the requirements of this standard. Employees who may be exposed are required to undergo a medical examination and acquire a hepatitis B vaccination. An example of personnel who may meet this requirement include plant medical personnel and first-aid responders, among others. Training requirements include a copy of the OSHA standard and explanation, in addition to information about the symptoms and epidemiology of bloodborne diseases; transmission of bloodborne pathogens; appropriate methods for recognizing jobs and other activities involving exposure to blood; use and limitations of practices that will prevent or reduce exposure, such as Engineering Controls and personal protective equipment; types, proper use, location, removal, handling, decontamination, and/or disposal of personal protective equipment; selecting personal protective equipment; hepatitis B vaccine, including information on its efficacy, safety, and benefits; appropriate actions to be taken and persons to contact in an emergency; procedure to be followed if an exposure incident occurs, including the method of reporting, the medical follow-up, and medical counseling; and signs and labels. Additional training may be required in applicable laboratory situations and other circumstances.
10. The employer is required to maintain complete and accurate medical records. These records must include the names and social security numbers of the employees; a record of the employee's hepatitis B vaccination and medical evaluation; the results of all physical examinations, medical testing, and follow-up procedures; the physician's written opinion; and all information provided to the physician. The employer is charged with the responsibility of maintaining the confidentiality of these records.
11. The employer is additionally required to maintain all training records, which must include dates of all training, names of persons conducting the training, and names of all participants. The training records must be maintained for a minimum of five (5) years. Employees and OSHA must be provided the ability to view and copy these records upon request.

12. All containers containing infectious waste, including but not limited to, refrigerators and freezers containing infectious waste, medical disposal containers, and all other containers, *must* be properly labeled. The required label must be fluorescent orange or orange-red with the appropriate symbol and the word "BIOHAZARD" in a contrasting color. This label *must* be affixed to all containers containing infectious waste and must remain affixed until the waste is properly disposed of.

*Note:* This important standard applies to all employers who have personnel assigned specifically to render first aid or other medical services.

# *chapter eleven*

# *Emergency and disaster preparedness*

*"Difficulties exist to be surmounted."*
—Ralph Waldo Emerson

*"The greatest difficulties lie where we are not looking for them."*
—Johann Wolfgang Von Goethe

Disasters occur somewhere in the world every day. Simply by watching the evening news or reading a newspaper we can readily find news of disasters hitting individuals, companies, and entire countries. Disasters take various forms, ranging from natural disasters such as tornadoes to manmade disasters such as workplace violence, which happens on a far too frequent basis. No matter what type of disaster, the results are typically the same — substantial loss of life, money, assets, and productivity.

Today's disaster risk has substantially evolved to include areas far beyond the natural disasters of the past, such as cyber-terrorism, product tampering, biological threats, and ecological terrorism, which were virtually unheard of a few short years ago. These disasters can be just as devastating to an organization as a natural disaster; however, the prevention and proactive measures taken are substantially different. Today's safety professional is faced with a myriad of new and different issues and reactions, ranging from control of the media to shareholder reaction, which were not even considered in the disaster preparedness programs of the past.

Because the world is changing, technology is changing, and risks are increasing and evolving, the safety profession must adapt in order to prevent potential disasters from happening where possible, minimize the risks where prevention is not possible, and appropriately react to keep damages to a minimum.

Appropriate planning and preparedness before a disaster occurs are essential to minimizing risks and resulting damages. Individuals involved in disaster preparedness efforts must be appropriately selected and trained so their responses and decision-making during a time of crisis are appropriate. Risks that cannot be appropriately managed internally should be assessed for the possibility of providing external protection or shifting the risk, by means of insurance, for example. Reaction after the disaster must be sure and coordinated in order to minimize damage and avoid causing further harm to the remaining assets, both tangible and intangible.

In this chapter, we have designed a new and innovative method for preparing companies and organizations to address the substantial risk of disasters in the workplace. Our methodology not only encompasses tried and true proactive methodology utilized by safety professionals for decades to address natural disasters but also addresses the often overlooked reactive and post-disaster phases. The progress of society and the accompanying technological changes and terrorist activities require that safety professionals re-think the standard *modus operandi* for disaster preparedness and expand their proactive and reactive measures to safeguard their company or organization's assets on all levels and at all times. In the blink of an eye, one disaster can decimate years of effort, creativity, and sweat, as well as the monetary and physical assets of a company or organization. Safety professionals are charged with the responsibility of identifying these risks, preparing to protect against these risks, and reacting properly if this risk should develop.

## Risk assessment

Risks of varying types and magnitudes exist in every workplace on a daily basis; however, some risks are far greater than others and can be disastrous if not identified and properly addressed in terms of reducing the probability of the risk (where feasible), protecting assets through shifting all or a percentage of the risk, and minimizing the potential harm of the risk in a disaster event. The initial step in any disaster preparedness endeavor is to identify the various potential risks, assess the viability and potential of the risk, evaluate the actual probability of the risk, and appraise the potential damage. The potential risks will vary from operation to operation, facility to facility, and location to location. It is vitally important that safety professionals properly identify and assess each facility and operation on an individual basis and customize the preparedness program and efforts to meet the needs of the unique facility. There is no one basic emergency and disaster plan that fits all facilities and operations.

## *Planning design*

Safety professionals should consider a plan that addresses all potential risks and provides the maximum protections for their employees. The plan should include, but not be limited to, the following considerations:

1. Evacuation routes — internal and external
2. Communication systems
3. Command areas
4. Specific responsibilities
5. Triage areas
6. Press areas
7. Liaisons with public sector assistance organizations (i.e., fire, police, EMS)
8. Liaisons with local medical facilities
9. Salvage and security

## *Media relations*

One essential area often overlooked in the preparation of an emergency and disaster preparedness plan is control of the information and image of your organization being projected to the world by the media. Preplanning in regard to the who, what, when, where, and how of information flow is essential to ensure the accuracy of the information being disseminated about your company and the emergency situation it is facing, as well as creating a favorable opinion among the public about your company's handling of the situation.

Consider the following example. Suppose a publicly held company experiences an explosion that results in ten fatalities, a large number of injured workers, and extensive damage to the facility. After the company notifies the fire department, EMS, and local law enforcement, the local media (who is usually monitoring radio transmissions) dispatches a reporter or television crew to the scene. The television crew will be working under a deadline to provide videotape and information about the incident as quickly as possible. They will try to make the videotape as graphic as possible to interest the viewers and will seek remarks from bystanders, employees, firefighters, or whomever is available (the information provided by witnesses may or may not be correct; however, the television station, for legal reasons, can add a disclaimer at the end of the telecast).

The information gathered at the incident scene will very quickly be picked up by global television networks such as CNN and various newspapers and magazines, in addition to finding its way onto the Internet for worldwide distribution. The information may be slanted or otherwise "editorialized" and often, as when children pass a secret from one to the next, begins to be modified, expanded, and otherwise changed to enhance the story. The facts and truth of the situation are often lost in the shuffle.

Members of the public, sitting at home watching the news or reading the newspaper, will make a value judgment about the company. They may be stockholders, potential employees, or customers who are concerned about the effect this incident will have on trading of the company's stock, future employment with the company, or availability of the company's products. In essence, the information provided to the general public by the media contributes to the opinions formed by general public about a particular company or organization (whether good or bad) which will affect the public's interactions with the company or organization in the future.

Control of the flow of information after a disaster situation is essential and should be part of an overall comprehensive emergency and disaster preparedness plan. Controlling information about a disaster situation is just as important today as every other phase of the plan due to the potential long-term, downstream, detrimental effects of mishandling the flow of information. As with all other components of the emergency and disaster preparedness plan, appropriate attention should be paid to all areas of the information flow to ensure that a company is putting the best "spin" on an already bad situation. Remember, a bell once sounded cannot be un-rung!

As part of the overall emergency and disaster preparedness efforts, consider the following:

- Where will the media acquire their information?
- Will information be screened by legal counsel before providing to the media?
- Will information provided to the media be scripted or *ad lib*?
- Who will be providing information to the media?
- Is the person providing the information a good representative of your company?
- What is the background of this person?
- What image is projected by this person (e.g., dress, voice quality)?
- What does the media know about your company or organization besides the disaster situation?
- Will the media detrimentally affect the emergency efforts?
- What type of videotape footage will the media be able to acquire?
- What are possible timetables for the media (e.g., a television crew needing to leave the scene by 4:45 p.m. to make a 6:00 p.m. newscast)?
- Where will the media park their vehicles?

Control of information is essential after a disaster situation. Consider the following measures to address the various media-related issues posed above as part of the overall emergency and disaster preparedness plan:

- Identify an area in the parking lot or away from the flow of emergency traffic to which security or management should direct all media vehicles.
- Guide the media to appropriate areas to acquire video footage.

- Maintain security in the media area to prohibit media representatives from "wandering" into the emergency area.
- Designate a selected member of management as spokesperson for the company; no other members of management will talk with the media.
- Consider the dress, tone of voice, ability to remain calm, and other attributes when selecting a spokesperson.
- Provide the spokesperson with an appropriate platform, microphone, and backdrop (e.g., visible company logo) from which to provide information to the media.
- Distribute informational packets about the company to the media.
- Screen all information by legal counsel prior to presentation and keep questions from the media to a minimum.
- Keep in mind the media's deadlines and, if possible, provide information in a timely manner; if information is not provided by the company, hearsay and file footage could be used.
- Always provide truthful information or no information at all.
- Remember the families of the injured or killed employees; do not release names prior to notification of next of kin.
- Remember that the company will be news only for a short period of time. There will be a lot of news potential at the facility or operation immediately after a disaster, but, while the story may be front page today, it could be last page tomorrow. It is critical to control the information flow while the situation is most intense.

## Remember the shareholder

Safety professionals should also consider the damage inflicted by a disaster on shareholder perceptions of the operations (and, potentially, stock value) of a publicly held company. Companies have become very bottom-line oriented, and the influence of Wall Street on day-to-day operations is significant. A disaster of any magnitude will be immediately communicated by the media throughout the world. Publicly held companies are owned by the individuals and entities holding their stock, and the worth of such a company is intrinsically tied to the value of its shares. If shareholders perceive that a disaster will have a negative impact on the value of their stock, they may decide to sell the stock. If a sufficient number of shares are sold, the value of the shares diminishes which leads to a lower value for the company which equates to tighter budgets, fewer personnel, and other subsequent impacts.

Safety professionals should be aware of how Wall Street can have a positive or negative impact on important aspects of their disaster preparedness program, such as budgets and staffing. Additionally, employees of publicly held companies participate in 401K plans, stock purchasing plans, stock option plans, and other retirement programs and are very interested in the value of their company's stock. Let's examine this issue a little more closely. If a company is "public," the actual owners of the company are the shareholders. The company's board of directors actually works for the shareholders, while the chief executive officer and other upper management officials report to the board of directors. The value of a publicly held company is based upon, in whole or in part, the value of its stock, which is traded daily on one of the stock exchanges (e.g., New York Stock Exchange, American Stock Exchange, NASDAQ).

Today's technology allows individuals to "track" individual stocks on a daily basis on television (e.g., MSNBC) and through Internet services. Additionally, more individual shareholders are managing their portfolios through online trading services, such as E-trade and Ameritrade. In the past, investors may have placed their money in mutual funds and checked the progress once or twice a year, but today they are monitoring the performance of individual company stocks on a daily, or even minute-by-minute, basis. In the past, investors may have "weathered the storm," but today they are more actively involved and more willing to purchase or sell individual stocks at a moment's notice.

So, how does this directly and indirectly affect the overall function of a company? Upper management is now more focused on quarterly returns and looking for a return on investment (ROI) for their shareholders. With such a focus, the safety function has been placed under a microscope to take steps to minimize losses, which may affect the bottom line. The safety manager may be required to report to upper management more often and to collect a wider range of data on a regular basis. Also, given the instantaneous communication available today, a serious accident, chemical release, or other significant event will make the news immediately and could have a detrimental effect upon the individual stock price (for example, the effect of the Valdez disaster on Exxon stock).

When a company is not doing well and shareholders are "jumping ship," the dollars available for safety programs may be reduced. This could have a detrimental effect on management of an emergency situation. Shareholders hold a company's stock to make money; some may hold onto the stock looking for long-run gains, but the current trend is for shareholders to hold stocks for only a short period before trading them. A major disaster situation can quickly place a company's stock into play, with shareholders having to decide whether or not to hold or sell the stock. If the situation appears to be appropriately managed, there is a higher likelihood that shareholders will "weather the storm." If shareholders perceive the situation as being poorly managed, though, they may decide to sell their stock.

In essence, shareholders become a company's bosses. Shareholders normally vote with their feet; if a company experiences a major accident, misses a dividend, or does not meet Wall Street's expectations, there is a substantial likelihood that shareholders will sell off the company's stock. When that happens, the value of the company tends to fall, which affects the company's borrowing power, overall operations, and effectiveness of the safety function.

The key in an emergency situation is to control the flow and type of information being disseminated to the general public. It is important, then, to have a plan established to give shareholders information regarding management of the emergency and how the emergency could affect their investment in the company. The control of information must be managed just as effectively as all other phases of an emergency and disaster plan.

In summary, below are several items to address within an overall disaster preparedness program:

- How will information be provided to the media?
- How will the media be controlled on site?
- Who will act as spokesperson for the company?
- How will the spokesperson be dressed?
- What backdrop will be utilized for videotape and photographs?
- What brokerage house brought the company's stock public?
- What brokerage house represents most of the company's shareholders?
- What lines of communications have been established with these brokerage houses?
- How quickly can information be provided to shareholders?
- Does the information being provided to shareholders reflect a confident posture?
- What type of follow-up should be provided to shareholders?
- Does the company maintain an Internet site?
- What information will the company provide on the Internet?
- Is the information being provided truthful and complete?
- Is the follow-up timely?

In today's market, it is easier for shareholders to sell stock in a company than to risk losses due to a disaster situation. It is vitally important that careful thought and preparation are given to the image the company will be projecting to the world in 30-second sound bytes. Shareholders, just like the general public, can make value judgments about a company from a tiny snippet of information. Appropriate management, preparation of image, and effective communication can minimize the overall and long-term damage to an organization. Safety professionals, then, should address every potential risk when developing their emergency and disaster preparedness programs. Every potential issue — ranging from such traditional elements as evacuation routes and triage areas to newer considerations such as media and shareholder information — should be addressed during the planning stage. Selected individuals who will serve in key capacities within the overall plan should be properly trained and prepared to function under stressful conditions if a disaster situation should arise. Preparation is the key to weathering a disaster situation successfully and minimizing the damage in terms of human and economic impact.

# *chapter twelve*

# *Chemical hazards*

*"Who is an honest man?*
*He that doest still and strongly good pursue.*
*To God, his neighbor and himself most true:*
*Whom neither force nor fawning can*
*Unpin, wrench from giving all their due."*
—Herbert

*"Honor and shame from no condition rise;*
*Act well your part; there all the honor lies."*
—Pope

## *Employee Right-To-Know standard*

### *History of chemicals in the workplace*

The Occupational Safety and Health Administration (OSHA) has estimated that more than 30 million American workers are exposed to hazardous chemicals in 3.5 million workplaces (48 FR 53282, 53323; 52 FR 31871). The National Institute for Occupational Safety and Health (NIOSH) reports that there are 375,000 hazardous chemicals in these workplaces and, due to the growth of the chemical industry, this number is rising. The *Federal Register* reports that chemical exposures occur in nearly all types of industry (52 FR 31858) and workers may experience exposure to multiple chemicals many times at one point in their career or over a long period of employment (48 FR 53323). OSHA believes it is every employee's right to know specific information about the hazardous chemicals in their workplace to be able to protect themselves from health problems related to exposure to those chemicals. Exposure to some chemicals may cause only mild irritation, while others may cause severe health problems or even death. Still other chemicals have been linked to chronic illnesses such as heart disease, kidney disease, sterility, and cancer. Some hazardous chemicals may pose physical hazards to the worker by initiating fires or explosions. Some chemicals upon contact will burn the skin or respiratory passages. The vapors from some chemicals are toxic in their normal state, and many other chemicals produce toxic byproducts if they are exposed or involved in a fire.

The Occupational Safety and Health Administration initiated data collection in preparation for developing a standard for communicating hazard information to the employees. The following information is reprinted from the *Federal Register* concerning their findings:

- During the Hazard Communication standard rulemaking, data collected about chemical illness and injury rates in manufacturing sectors showed that some 40–50,000 manufacturing workers experienced chemical-source illnesses each year, and an average of 10,000 workers' compensation claims were filed annually in connection with chemical illness or injury in manufacturing (48 FR 53285).
- Employees in non-manufacturing industries were estimated to experience acute chemical illness and injury at the rate of 13,671 injuries, 38,248 illnesses, and 102 fatalities per year (52 FR 31868).
- The chronic disease rate was found to be 17,153 chronic illnesses, 25,388 cancer cases, and 12,890 cancer deaths per year.

The Occupational Safety and Health Administration believes that the reported data do not accurately reflect the severity of the health and safety problems caused by chemicals in the workplace. OSHA also thinks a lack of knowledge about health effects associated with chemical exposures contributes to the chronic under-reporting of occupational illnesses. Because many of the health-related problems are also diseases or physical problems that may occur in workers as a result of non-chemical or non-occupational exposure, it is often difficult to determine that such ailments are the direct cause of occupational exposures. Misdiagnosis is a problem, and often symptoms are treated without realizing that the cause is an occupational chemical exposure (53 FR 25973).

Employee turnover also causes some chemical exposure problems to go unreported. Exposure to some chemicals may not manifest itself in symptoms for as long as 20 to 30 years. This is especially true for some chronic illnesses. The employee may never make the connection between exposure to a chemical 20 or more years ago and current health problems. Chemical exposure can also reduce the capacity of employees to perform their duties safely. A fall from a ladder or amputation may not be linked to a chemical exposure that has diminished the employee's judgment or physical ability. Based upon the data collected, the rulemaking process began, and today workers are protected by 29 C.F.R. 1910.1200, the Hazard Communication standard, also known as the Employee Right-To-Know standard.

## *Overview of the Hazard Communication standard (29 C.F.R. 1910.1200)*

The Hazard Communication standard ensures that the hazards of all chemicals produced or imported are evaluated and that information concerning their hazards is made available to employers and employees. The Hazard Communication standard involves three separate but integrated requirements:

1. Chemical manufacturers and importers must review available scientific evidence concerning the physical and health hazards of the chemicals they produce or import to determine if they are hazardous [paragraph (d)].
2. For every chemical found to be hazardous, the chemical manufacturer or importer must develop comprehensive Material Safety Data Sheets (MSDSs) and warning labels for containers and send both downstream along with the chemicals [paragraphs (f), (g)].
3. All employers must develop a written hazard communication program and provide information and training to employees about the hazardous chemicals in their workplace [paragraphs (e), (h)].

Container and package labels, Material Safety Data Sheets, and worker training combine to educate both the employers and the employees and motivate them to modify their behaviors and minimize exposure to these hazardous chemicals. Armed with information from Material Safety Data Sheets about the physical and health hazard properties of the chemicals they work with, employees are better prepared to recognize dangers and take appropriate countermeasures. The individuals responsible for educating the workers must realize it is not enough to enable the employees to recite all the hazards associated with a particular chemical. It serves no purpose to educate the employees on the flash point of a flammable liquid or the permissible exposure limit (PEL) of a toxic chemical if the employee does not understand how that information relates to safe use and handling of that chemical. The real purpose in conducting the Hazard Communication standard training is to ensure that the employees trained:

1. Recognize all the hazards associated with all the chemicals they work with so they are motivated to work safely.
2. Can demonstrate how to work safely to prevent exposure to the chemicals which may involve any or all of the following:
   - Administrative controls
   - Engineering controls
   - Personal protective equipment (PPE)
3. Demonstrate they can make wise choices to minimize the negative impact of any unexpected exposure to hazardous chemicals, including:

- Emergency procedures such as evacuation
- Emergency notification procedures
- First-aid techniques

The comprehensive technical information provided on each MSDS also serves as a reference document for exposed workers and healthcare professionals providing services to those workers. Labels required by the Hazard Communication standard should provide instant danger recognition and jog an employee's memory concerning the hazards associated with each container being handled. The standard was promulgated to modify both employer and employee behavior and minimize the number of hazardous chemical contact incidents. It also aids employers in purchasing effective PPE and motivates employees to wear it and follow safe work practices.

Although the requirements in this standard are relatively simple, implementing it has proved difficult for some employers. Since OSHA began enforcing the standard, it has remained at or near the top of the list of citations issued for violations of OSHA regulations.

### *Standards cited for all SICs*

Listed below are the top ten standards cited by OSHA for all SIC (Standard Industrial Classification) codes during the period October 1998 through September 1999. Penalties shown reflect current rather than initial amounts.

| Standard | No. Cited | No. Inspected | Penalty ($) | Description |
|---|---|---|---|---|
| 1910.1200 | 7122 | 3642 | 1,563,656.81 | Hazard communication |
| 1926.0451 | 6723 | 2515 | 7,079,110.05 | General requirements for all types of scaffolding |
| 1926.0501 | 4137 | 3462 | 6,866,322.83 | Fall protection scope/applications/definitions |
| 1910.0147 | 3886 | 2064 | 5,472,655.33 | Control of hazardous energy, lockout/tagout |
| 1910.0134 | 3742 | 1590 | 1,110,067.31 | Respiratory protection |
| 1910.0305 | 3193 | 1866 | 1,220,003.20 | Electrical, wiring methods, components, and equipment |
| 1910.0212 | 2840 | 2201 | 4,211,889.19 | Machines, general requirements |
| 1910.0219 | 2445 | 1222 | 1,514,841.28 | Mechanical power-transmission apparatus |
| 1910.0303 | 2231 | 1574 | 1,394,088.50 | Electrical systems design, general requirements |
| 1910.0217 | 2102 | 598 | 1,540,594.97 | Mechanical power presses |

*Source:* http://www.osha.gov/oshstats/std1.html.

Included in complaints by employers about compliance with the Hazard Communication standard are

- Difficulty in obtaining the MSDSs in a timely manner
- Problems associated with managing the ever-changing database of MSDSs; only the MSDSs for the chemicals currently in the work area(s) are to be kept in the active file, but MSDSs can be maintained in an inactive file if the chemical is no longer in the work area(s)
- Difficulty in properly documenting the written program
- Problems associated with employee failure to properly label temporary containers

## *Community right-to-know: a history of mistrust and disasters*

### *Emergency Planning and Community Right-to-Know Act (42 U.S.C. 11001 et seq., 1986)*

Also known as Title III of SARA (Superfund Amendment Re-Authorization Act), the Emergency Planning and Community Right-To-Know Act (EPCRA) was enacted by Congress as the national legislation on community safety. This law was promulgated to help local communities protect public health and safety and the environment from chemical hazards. The law was passed by Congress in response to a growing national sentiment that large industrial companies were polluting the land, sea, and air and, at the same time, failing to inform the citizens who lived near these industrial plants of the dangers they faced. Many people feel that one incident in December, 1984, became the catalyst for action. A pesticide factory owned by Union Carbide in Bhopal, India, exploded, killing 2500 people and injuring an additional 300,000. With a population of 700,000 people, Bhopal is the capital of Madhya Pradesh, one of India's poorest and least developed states. The city is geographically divided between rich and poor sections, with the factory located in the poor section. Although the company was a multinational, Indian investors owned almost half of the shares of the Indian plant, and Indians operated the plant.

The active ingredient for the pesticide, methyl isocyanate, was stored in 600-gallon tanks. The size of the tanks was a problem. Larger tanks are economically more efficient, as they can hold more gas, but they pose greater risks in the event of a tank leak. Regulations in Germany required a similar Union Carbide plant in that company to restrict its tank size to 100 gallons. The tank that exploded in the Indian plant was supposed to be refrigerated to 0°C; instead, the refrigeration unit was not working and the tank was at room temperature. Although the Indian factory had safety features to prevent disasters, several of the safety systems were not functioning. The temperature alarm was shut down; the gas scrubber, which was supposed to neutralize escaped gas, was shut off; and a flare tower, which was supposed to burn escaped gas, was out of service.

The explosion started when water was added to a 600-gallon tank of the chemical, possibly done as an act of sabotage by a disgruntled employee. A chain reaction led to the temperature in the tank rising, and the tank blew up. A fog of the gas drifted through the streets of Bhopal, killing people where they stood. Long-term medical problems for the survivors included respiratory ailments and neurological damage. The Indian government quickly arrested plant managers and eventually spent $40 million on various disaster-relief projects. Union Carbide stock plummeted, with losses totaling almost $1 billion; Union Carbide sales were also impacted for several years. The company eventually paid nearly $.5 billion dollars to the Indian government for the victims. Although the U.S. parent company reacted quickly and compassionately to the disaster, the tragedy raised serious questions about the parent company's views on safety and further eroded the confidence and trust of the American people. In 1986, Congress passed the Emergency Planning and Community Right-To-Know Act (SARA Title III).

## *Emergency Planning and Community Right-To-Know Act mandates*

To implement the EPCRA, Congress required each state to appoint a State Emergency Response Commission (SERC). The SERCs were required to divide their states into emergency planning districts and to name a local emergency planning committee (LEPC) for each district. The legislation mandated broad representation by firefighters, health officials, government and media representatives, community groups, industrial facilities, and emergency managers to ensure that all necessary elements of the planning process were represented. This also gave consumers confidence in the legislation because they felt the media would not suppress negative corporate environmental stories. For specific details concerning this federal law, consult the appropriate section of Title 42, chapter 116. An outline of the requirements of EPCRA by section follows.

### *Subchapter I — emergency planning and notification*

1. §11001. Establishment of state commissions, planning districts, and local committees
   - (a) Establishment of state emergency response commissions
   - (b) Establishment of emergency planning districts
   - (c) Establishment of local emergency planning committees
   - (d) Revisions
2. §11002. Substances and facilities covered and notification
   - (a) Substances covered
   - (b) Facilities covered
   - (c) Emergency planning notification
   - (d) Notification of administrator

3. §11003. Comprehensive emergency response plans
   (a) Plan required
   (b) Resources
   (c) Plan provisions
   (d) Providing of information
   (e) Review by state emergency response commission
   (f) Guidance documents
   (g) Review of plans by regional response teams
4. §11004. Emergency notification
   (a) Types of releases
   (b) Notification
   (c) Follow-up emergency notice
   (d) Transportation exemption not applicable
5. §11005. Emergency training and review of emergency systems
   (a) Emergency training
   (b) Review of emergency systems

*Subchapter II — reporting requirements*

1. §11021. Material Safety Data Sheets
   (a) Basic requirement
   (b) Thresholds
   (c) Availability of MSDS on request
   (d) Initial submission and updating
   (e) "Hazardous chemical" defined
2. §11022. Emergency and hazardous chemical inventory forms
   (a) Basic requirement
   (b) Thresholds
   (c) Hazardous chemicals covered
   (d) Contents of form
   (e) Availability of tier II information
   (f) Fire department access
   (g) Format of forms
3. §11023. Toxic chemical release forms
   (a) Basic requirement
   (b) Covered owners and operators of facilities
   (c) Toxic chemicals covered
   (d) Revisions by administrator
   (e) Petitions
   (f) Threshold for reporting
   (g) Form
   (h) Use of release form
   (i) Modifications in reporting frequency
   (j) EPA management of data
   (k) Report
   (l) Mass balance study

*Subchapter III — general provisions*

1. Relationship to other law
   - (a) In general
   - (b) Effect on MSDS requirements
2. §11042. Trade secrets
   - (a) Authority to withhold information
   - (b) Trade secret factors
   - (c) Trade secret regulations
   - (d) Petition for review
   - (e) Exception for information provided to healthcare professionals
   - (f) Providing information to administrator; availability to public
   - (g) Information provided to state
   - (h) Information on adverse effects
   - (i) Information provided to Congress
3. §11043. Provision of information to healthcare professionals, doctors, and nurses
   - (a) Diagnosis or treatment by healthcare professional
   - (b) Medical emergency
   - (c) Preventive measures by local healthcare professionals
   - (d) Confidentiality agreement
   - (e) Regulations
4. §11044. Public availability of plans, data sheets, forms, and follow-up notices
   - (a) Availability to public
   - (b) Notice of public availability
5. §11045. Enforcement
   - (a) Civil penalties for emergency planning
   - (b) Civil, administrative, and criminal penalties for emergency notification
   - (c) Civil and administrative penalties for reporting requirements
   - (d) Civil, administrative, and criminal penalties with respect to trade secrets
   - (e) Special enforcement provisions for §11043
   - (f) Procedures for administrative penalties
6. §11046. Civil actions
   - (a) Authority to bring civil actions
   - (b) Venue
   - (c) Relief
   - (d) Notice
   - (e) Limitation
   - (f) Costs
   - (g) Other rights
   - (h) Intervention
7. §11047. Exemption
8. §11048. Regulations
9. §11049. Definitions
10. §11050. Authorization of appropriations

# chapter thirteen

# Workplace violence

*"Nothing good ever comes of violence."*
—Martin Luther King

*"Violence is the last refuge of the incompetent."*
—Isaac Asimov

Violence in the workplace has quickly become a major consideration for safety professionals in order to safeguard their employees and organizational assets. In recent years, incidents of workplace violence, ranging from physical altercations to firearms use, have increased significantly. Proper preparedness to address the potential risk of violence in the workplace is a major area of concern for the Occupational Safety and Health Administration (OSHA) and, thus, for safety professionals, who should consider taking any and all precautions necessary to keep their employees safe. Remember that employees want to return home from work at the end of the day in the same condition as when they went in to work. It is the safety professional's job to ensure that this happens.

## Individual warning signs

Several studies have found that some individuals are prone to violence in the workplace. Safety professionals should be able to identify the signs of potential violence and work to provide methods to address the situation before the employee becomes violent. An individual that commits workplace violence does not have a sign on his back that says, "I will commit a violent act in the workplace." That, of course, would be too simple; instead, the signs are subtle, and even the person himself does not know that he will commit a violent act one day. One event can start that first domino falling and it is difficult to stop all the dominoes from falling once they have started.

An individual who will commit a violent act in the workplace does not fit a particular profile; there are no clear-cut categories. Many experts have carried out research into workplace violence and have found some broad categories that violent employees fit into. Such employees will have worked at their place of employment for a while and had some success in their jobs. They are dedicated employees who come to work on Saturdays or stay late to finish a project. They make work their life; their work determines how they view themselves, and everything in their life is focused around their work schedule.

Then, one day, they may feel that they were passed up for a promotion or raise and may start holding a grudge against a certain person or the company itself. Another scenario is that the company is downsizing and some people have to be laid off or the company was bought by another one and there will be layoffs. A grudge may be held against a certain person whom they think is responsible for the lack of a raise or promotion or for being laid off.

These people may begin to think that someone is out to get them — whoever laid them off or got the promotion instead of them. They may also be having a problem outside of work that they are trying to deal with. It could be with their spouse, family, or an annoying neighbor, but they are always having a problem and are always talking about it. They may also have a history of violence, whether a fight in a bar or spousal abuse. They may be fascinated with guns, owning several and even carrying a concealed one.

Some experts say that the typical person prone to workplace violence is a white male in his thirties or forties. As discussed previously, he would have worked for his company for a long period of time, would have a history of violence, and may own a firearm. He is most likely a loner and may or may not have a family; if he does have a family, they will be a source of problems.

Such an employee may try to intimidate the person against whom he holds a grudge by making veiled threats or stalking them. Or, the person may show no outward visible signs of hostility. Take the case of Larry, a technician for an electronics manufacturer in California. Larry was one of several employees laid off due to the recession in 1991. After he was laid off, he kept returning to his former place of employment, seeking advice about finding another job or just visiting with people. Then one day he walked in with a bandoleer across his chest and a shotgun in his hand. He shot out the switchboard before the receptionist could call for help. After setting small fires and setting off several pipe bombs, he then calmly walked upstairs and shot and killed a vice president and a regional sales manager. Another executive escaped with his life by hiding under a desk. When Larry was done, he walked out and got on his bicycle and rode away.*

* Barrier, M., The enemy within, *Nation's Business*, Feb., pp. 18–22, 1995.

There is no distinct category of people who commit violent acts, but there are warning signals, such as those that were mentioned previously. Noticing some of these warning signals should heighten concern among an employer. An employee displaying some of these signs does not mean he will perform a violent act, but the company should have a prevention program in place so when the signs are recognized the program can be put into practice.

The following list includes some of the warning signs for potential for violence in the workplace:

- The person holds grudges against coworkers or the company itself; he might not have received a promotion or a raise that he felt was deserved.
- The person exhibits paranoid behavior; he may feel that the whole world is out to get him.
- The person regularly has problems, which may be related to home, work, or outside friends; the person talks frequently about the problem he is currently dealing with.
- The person has a fascination with firearms and may carry one on his person; he may have a collection and frequently talks about them.
- The person suspects that he is going to be fired or laid off.
- The person is a loner at work and does not mix with coworkers, except maybe for a romantic interest that may lead to the coworker feeling threatened.
- The person may intimidate others by verbal or physical means (e.g., anonymous notes that threaten someone or harassing phone calls).
- The person has difficulty accepting criticism and/or authority.
- The person likes to push authority to its limit, to see how far he can go.
- The person has a history of violent behavior, either before or after employment with the company.
- The person has a history of substance abuse, either drugs or alcohol, or both.
- The person is fascinated by acts of workplace violence and likes to talk about them with coworkers.
- The person may come from an unstable or dysfunctional family.
- Workplace events, such as downsizing or a corporate takeover, may generate stress in the person.
- Activities that expose the person to outside factors may cause an act of violence.*

## *Workplace violence prevention program*

The Occupational Safety and Health Administration has stated in their regulations [29 U.S.C. §654(a)(1)] that "any employer in the business to provide goods and services that employs people shall provide a safe and healthful

* Barrier, M., Johnson, D., Kiehlbaugh, J., and Kurutz, J., Workplace violence scenario for supervisors, *Human Resources Magazine*, pp. 63–67, 1995.

work environment." This means that when we go to work we should not have to worry about someone bothering us or harassing us. But that is not always the case anymore, as a disgruntled worker at any time can take out his misguided anger on coworkers or the company he works for. Such violence has escalated to the point where it is almost out of hand; if companies do not take the initiative to stop it, there is no telling where it could lead.

We all like to count on our workplace being violence free. Civility and cooperation are basic and essential to the work environment, but now violence is not just on the streets; it has ventured into the office and we must all confront it. In the past, workers tried to leave their home lives at home. We can all remember a boss saying, "Don't bring your personal life to the office." Our world is becoming more violent, and more people are bringing that violence to work with them rather than leaving it at the door. When it happens, it is a rude awakening, as most people think that such violence couldn't happen to them or their employers; it could only happen to someone else.

Many companies have realized what is happening and are taking preventive measures to guard against it. Employee assistance programs (EAPs), which originally started with World War II to deal with alcoholism, are being used to deal with employees who have such problems as marital difficulties or substance abuse. Such programs foster loyalty to a company and a more productive workforce, in addition to assuring employees that the company cares about them, which can sometimes mean more than a raise. Most EAPs are run by outside agencies, as rarely are there people in the company trained to deal with these types of problems, plus confidentiality is ensured.

A preventive maintenance plan for violence should be in effect to deal with violence or even the threat of violence. The following is an outline of a preventive program for a company:

1. *Create a management team to develop, review, and implement policies dealing with violence in the workplace.* These policies should be included with the company's other policies and procedures. Management backing is very important to any type of employment program. When creating the management team, each team member should be assigned a well-defined responsibility. Once the team has been created, the first step should be an assessment of the possibility of a violent act occurring in their workplace. The team members should make themselves experts on the subject of workplace violence. They should contact all types of outside resources so as to better equip themselves on the subject. The team should then prepare for the possibility of a violent act occurring in their workplace by developing an action plan to deal with a violent act and educating the rest of the employees, supervisors, or team leaders.
2. *Create a program that helps employees to detect early warning signals that could ultimately lead to a violent act.* This program will enable supervisors and/or mangers to detect early warning signals and deal with them head on before violence erupts. This program can be incorporated into

an already existing training program or be provided as a separate program. Managers, supervisors, or team leaders should be trained to spot the signs, including those discussed previously, and to report them to a member of the management team. The appropriate member of the management team should then conduct an investigation. Any downsizing in the company should put supervisors especially on the alert for any signs of potentially violent behavior.

3. *Develop an investigation format for dealing with complaints about a potentially violent employee.* The first rule regarding a threat is to take it seriously, as any threat could lead to a violent act. Investigating and recording incidents are key to preventing a violent act in the workplace. Such investigations should include these four Ws:

   - *Who:* Who made the actual complaint or threat? Who witnessed the actual threat?
   - *When:* When did the threat take place? Are there any events leading up to the actual threat; if so, when did they take place?
   - *What:* What exactly happened? Include all facts that pertain to the actual threat and the investigator's assessment of what occurred.
   - *Where:* Where did the threat take place? In an office or outside the company? (This is an important aspect that easily gets left out.)

   An incident report form already prepared by the investigator would be the easiest way to handle the investigation. Questions on the report form may include:

   - Who made the threat?
   - Who was the threat made against?
   - What were the actual words spoken in the threat?
   - Was there any threat of physical harm? If so, what specifically?
   - Where did the threat take place (actual physical location)?
   - What time did the threat take place (be specific)?
   - Who witnessed the actual threat?
   - What did the witnesses say occurred (in their own words)? Ask witnesses to write down their statements and attach their statements to the document.
   - Is there any other information that might pertain to the incident?

   These questions are not inclusive, and a company needs to make their own incident form specific to their organization. It is important when interviewing the person who made the threat to do so in a nonthreatening environment. Having a strategy to deal with the person's anger is important. Focusing only on the facts, not on personal opinions or personality traits, will help the person deal with the situation in a calm, rational manner and control emotional reactions. In addition, there are steps that the interviewer can take to let the person know that the interviewer is interested in what is being said, such as making eye contact; paying full attention to what the person is saying (by avoiding working on notes or answering the phone, as doing so can make the situation worse); asking open-ended questions; not interrupting the person; letting the person finish what he is saying; speaking in a calm, rational voice; and repeating what the person says to clarify the situation.

4. *Take any action necessary that can prevent an incident from occurring.* This may be disciplinary action, talking to a member of the management team in charge of workplace violence, or referral to the company's employee assistance program for counseling. The investigator interviewing the person will come to a set of conclusions. The action that is taken should be based on the conclusions drawn from the interview. In some cases, the person who made the threats may feel better just after talking to the interviewer. This may resolve the issue, but in most cases it won't and further action will be necessary. It is important to take a zero-tolerance policy toward any threat. Consider a company that disciplines someone who makes a threat by making him take a day off work without pay. When this same person later hits another employee, he is made to take three days off without pay. This type of action will not help the situation; it will only make it worse. The idea is to help an adult solve a problem rather than treat him like a naughty child that needs to be disciplined.
5. *Develop or improve upon security measures.* Technology has advanced security measures. Gone are the days of the security guard walking a solitary route around a building. Instead, we now have security cameras and laser beams to detect intruders. Electronic systems can provide or deny access to persons entering a workplace. Every employer needs to assess their needs and provide adequate security measures. Security companies will provide a review and recommend security measures, some of which will handle every aspect of installing and maintaining the security measures. Providing security is key to protecting employees from the threat of a violent employee and from other types of incidents that could threaten a company. Immediate improvements that can provide a measure of safety include outdoor lighting, closed-circuit television, intercom systems, and an alarm system. In addition to physical security measures, policies concerning security measures need to be included within the corporate policies and procedures manual. What does the policy state about an employee wanting to work late? Does another employee need to stay also? Are employees even allowed to stay after hours? The policy also needs to address security when a threat has been made. Is security tightened? If so, how? Local law enforcement agencies can be contacted for valuable information regarding threats occurring to other companies, mistakes that were made, and lessons learned from these mistakes. A mistake learned from one company may save a life in another. Local law enforcement can be a vital source of information, if companies utilize them.

6. *Develop a crises plan in the event of an actual violent act in the workplace.* The chain of command in the management team needs to be detailed so everyone understands his or her individual responsibilities and there is not any confusion when an incident occurs. When an incident occurs, responsibility shifts from the office manager or supervisor to a management team member, and these responsibilities need to be detailed. When are other agencies notified? The local police should be the first agency notified. A person from the employee assistance program needs to be notified and on hand. Other agencies should be notified as necessary, such as the fire department and local emergency services. In the crisis plan, other resources should be made available, such as a trauma consultant, either a counselor or therapist to assist employees in dealing with the crisis, a security consultant, an in-house legal representative, and a medical physician. What procedures should be put into place immediately to determine how safe the workplace is? Is there an internal security team that can isolate the area until the police department arrives? An immediate assessment is necessary to determine the effect the incident has had on the workplace and an investigation into the incident should be initiated:

   - What are the facts of the case?
   - Who were the witnesses?
   - How can the investigation the employer is conducting be coordinated with the investigation the police department will conduct?
   - What limits should the employer place on the release of information to other employees? Procedures about releasing information to other employees should be in place, as well as regulations about what can be said during emergency situations.
   - What are the immediate public relations concerns? How much information will be released to the public and who will release the information?
   - How does a company preserve control of the day-to-day running of the business while an incident is occurring?

   These are just a few of the considerations to be covered by a crisis plan. Each company needs to develop a plan specific to the organization. After a plan has been written, a hypothetical scenario should be enacted to let the employees know that the company is serious about dealing with workplace violence.
7. *Use the courts to deal with threats of violence.* The company needs to have a zero-tolerance policy toward threats of workplace violence or an actual incident of violence. When an employee has been discharged or is no longer employed by the company and makes a threat against the company or a representative of the company, the use of the court system to deal with the threat can be effective. Restraining orders prohibit a person from seeking access to the person or persons they have threatened. This is not a guarantee that an incident will be

prevented, but it is a preventive measure that will lessen the threat of an attack. When a state allows a restraining order, the person making the threat has to stay a certain distance away from the person they made a threat against. Such a restraining order gives the police the authority to make an arrest if the restraining order is violated.

In summation, violence in the workplace is not a traditional area of concern for most safety professionals; however, given the increase in incidents, the potential risk to the organization, and the expertise and skill level of the safety professional, this has become a major area of responsibility for safety professionals. Proper preparation, "thinking outside of the box," and addressing situations before violence occurs can safeguard an organization from the risk of violence. Additionally, proper preparation can minimize the potential damage if an incident of workplace violence occurs. Violence in the workplace is on the rise ... safety professionals must be prepared.

# *Appendixes*

# *appendix A*

# *OSHA inspection checklist**

The following is a recommended checklist to help safety and resource control professionals prepare for an Occupational Safety and Health Administration (OSHA) inspection.

1. Assemble a team from the management group and identify specific responsibilities in writing for each team member. The team members should be given appropriate training and education and should include, but not be limited to:
   a. OSHA inspection team coordinator
   b. Document-control individual
   c. Individuals to accompany the OSHA inspector
   d. Media coordinator
   e. Accident investigation team leader (where applicable)
   f. Notification person
   g. Legal advisor (where applicable)
   h. Law enforcement coordinator (where applicable)
   i. Photographer
   j. Industrial hygienist
2. Decide on and develop a company policy and procedures to provide guidance to the OSHA inspection team.
3. Prepare an OSHA inspection kit, including all equipment necessary to properly document all phases of the inspection. The kit should include equipment such as a camera (with extra film and batteries), a tape player (with extra batteries), a video camera, pads, pens, and other appropriate testing and sampling equipment (e.g., a noise-level meter, an air sampling kit).

* Schneid, T., Preparing for an OSHA inspection, *Kentucky Manufacturer,* Feb., 1992.

4. Prepare basic forms to be used by the inspection team members during and following the inspection.
5. When notified that an OSHA inspector has arrived, assemble the team members along with the inspection kit.
6. Identify the inspector. Check his or her credentials and determine the reason for and type of inspection to be conducted.
7. Confirm the reason for the inspection with the inspector (targeted, routine inspection, accident, or in response to a complaint?).
   a. For a random or target inspection:
      - Did the inspector check the OSHA 200 Form?
      - Was a warrant required?
   b. For an employee complaint inspection:
      - Does the inspector have a copy of the complaint? If so, obtain a copy.
      - Do allegations in the complaint describe an OSHA violation?
      - Was a warrant required?
      - Was the inspection protested in writing?
   c. For an accident investigation inspection:
      - How was OSHA notified of the accident?
      - Was a warrant required?
      - Was the inspection limited to the accident location?
   d. If a warrant is presented:
      - Were the terms of the warrant reviewed by local counsel?
      - Did the inspector follow the terms of the warrant?
      - Was a copy of the warrant acquired?
      - Was the inspection protested in writing?
8. The opening conference:
   a. Who was present?
   b. What was said?
   c. Was the conference taped or otherwise documented?
9. Records:
   a. What records were requested by the inspector?
   b. Did the document-control coordinator number the photocopies of the documents provided to the inspector?
   c. Did the document-control coordinator maintain a list of all photocopies provided to the inspector?
10. Facility inspection:
    a. What areas of the facility were inspected?
    b. What equipment was inspected?
    c. Which employees were interviewed?
    d. Who was the employee or union representative present during the inspection?
    e. Were all the remarks made by the inspector documented?
    f. Did the inspector take photographs?
    g. Did a team member take similar photographs?

# *appendix B*

# *Employee workplace rights**

## *Introduction*

The Occupational Safety and Health (OSH) Act of 1970 created the Occupational Safety and Health Administration (OSHA) within the Department of Labor and encouraged employers and employees to reduce workplace hazards and to implement safety and health programs. In so doing, this gave employees many new rights and responsibilities, including the right to do the following:

- Review copies of appropriate standards, rules, regulations, and requirements that the employer should have available at the workplace.
- Request information from the employer on safety and health hazards in the workplace, precautions that may be taken, and procedures to be followed if the employee is involved in an accident or is exposed to toxic substances.
- Have access to relevant employee exposure and medical records.
- Request the OSHA area director to conduct an inspection if he or she believes hazardous conditions or violations of standards exist in the workplace.
- Have an authorized employee representative accompany the OSHA compliance officer during the inspection tour.
- Respond to questions from the OSHA compliance officer, particularly if there is no authorized employee representative accompanying the compliance officer on the inspection "walkaround."
- Observe any monitoring or measuring of hazardous materials and see the resulting records, as specified under the OSH Act, and as required by OSHA standards.

* From http://www.pueblo.gsa.gov/cic_text/employ/osha/osha3021.html.

- Have an authorized representative, or themselves, review the Log and Summary of Occupational Injuries (OSHA No. 200) at a reasonable time and in a reasonable manner.
- Object to the abatement period set by OSHA for correcting any violation in the citation issued to the employer by writing to the OSHA area director within 15 working days from the date the employer receives the citation.
- Submit a written request to the National Institute for Occupational Safety and Health (NIOSH) for information on whether any substance in the workplace has potentially toxic effects in the concentration being used, and have their names withheld from the employer, if requested.
- Be notified by an employer if the employer applies for a variance from an OSHA standard, testify at a variance hearing, and appeal the final decision.
- Have their names withheld from their employer, upon request to OSHA, if they sign and file written complaints.
- Be advised of OSHA actions regarding a complaint and request an informal review of any decision not to inspect or to issue a citation.
- File a Section 11(c) discrimination complaint if punished for exercising the above rights or for refusing to work when faced with imminent danger of death or serious injury and there is insufficient time for OSHA to inspect; or file a Section 31105 reprisal complaint (under the Surface Transportation Assistance Act [STAA]).

Pursuant to Section 18 of the Act, states can develop and operate their own occupational safety and health programs under state plans approved and monitored by federal OSHA. States that assume responsibility for their own occupational safety and health programs must have provisions at least as effective as those of federal OSHA, including the protection of employee rights. There are currently 25 state plans. Twenty-one states and two territories administer plans covering both private and state and local government employment, and two states cover only the public sector. All the rights and responsibilities described in this booklet are similarly provided by state programs. Any interested person or groups of persons, including employees, who have a complaint concerning the operation or administration of a state plan may submit a Complaint About State Program Administration (CASPA) to the appropriate OSHA regional administrator. Under CASPA procedures, the OSHA regional administrator investigates these complaints and informs the state and the complainant of these findings. Corrective action is recommended when required.

## *OSHA standards and workplace hazards*

Before OSHA issues, amends, or deletes regulations, the agency publishes them in the *Federal Register* so that interested persons or groups may comment. The employer has a legal obligation to inform employees of OSHA safety and health standards that apply to their workplace. Upon request, the employer must make available copies of those standards and the OSHA law itself. If more information is needed about workplace hazards than the employer can supply, it can be obtained from the nearest OSHA area office.

Under the OSH Act, employers have a general duty to provide work and a workplace free from recognized hazards. Citations may be issued by OSHA when violations of standards are found and for violations of the general duty clause, even if no OSHA standard applies to the particular hazard. The employer also must display in a prominent place the official OSHA poster that describes rights and responsibilities under the OSH Act.

## *Right to know*

Employers must establish a written, comprehensive hazard communication program that includes provisions for container labeling, Material Safety Data Sheets, and an employee training program. The program must include a list of the hazardous chemicals in each work area, the means the employer uses to inform employees of the hazards of non-routine tasks (for example, the cleaning of reactor vessels), hazards associated with chemicals in unlabeled pipes, and the way the employer will inform other employers of the hazards to which their employees may be exposed.

## *Access to exposure and medical records*

Employers must inform employees of the existence, location, and availability of their medical and exposure records when employees first begin employment and at least annually thereafter. Employers also must provide these records to employees or their designated representatives, upon request. Whenever an employer plans to stop doing business and there is no successor employer to receive and maintain these records, the employer must notify employees of their right of access to records at least 3 months before the employer ceases to do business. OSHA standards require the employer to measure exposure to harmful substances; the employee (or representative) has the right to observe the testing and to examine the records of the results. If the exposure levels are above the limit set by the standard, the employer must tell employees what will be done to reduce the exposure.

## *Cooperative efforts to reduce hazards*

OSHA encourages employers and employees to work together to reduce hazards. Employees should discuss safety and health problems with the employer, other workers, and union representatives (if there is a union). Information on OSHA requirements can be obtained from the OSHA area office. If there is a state occupational safety and health program, similar information can be obtained from the state. OSHA provides special recognition through its Voluntary Protection Programs (VPPs) to worksites where employers and employees work together to achieve safety and health excellence.

### OSHA state consultation service

If an employer, with the cooperation of employees, is unable to find acceptable corrections for hazards in the workplace or if assistance is needed to identify hazards, employees should be sure the employer is aware of the OSHA-sponsored, state-delivered, free consultation service. This service is intended primarily for small employers in high-hazard industries. Employers can request a limited or comprehensive consultation visit by a consultant from the appropriate state consultation service.

## OSHA inspections

If a hazard is not being corrected, an employee should contact the OSHA area office (or state program office) having jurisdiction. If the employee submits a written complaint and the OSHA area or state office determines that there are reasonable grounds for believing that a violation or danger exists, the office conducts an inspection.

### Employee representative

Under Section 8(e), the workers' representative has a right to accompany an OSHA compliance officer (also referred to as a compliance safety and health officer, CSHO, or inspector) during an inspection. The representative must be chosen by the union (if there is one) or by the employees. Under no circumstances may the employer choose the workers' representative.

If employees are represented by more than one union, each union may choose a representative. Normally, the representative of each union will not accompany the inspector for the entire inspection but will join the inspection only when it reaches the area where those union members work.

An OSHA inspector may conduct a comprehensive inspection of the entire workplace or a partial inspection limited to certain areas or aspects of the operation.

### Helping the compliance officer

Workers have a right to talk privately to the compliance officer on a confidential basis whether or not a workers' representative has been chosen. Workers are encouraged to point out hazards, describe accidents or illnesses that resulted from those hazards, describe past worker complaints about hazards, and inform the inspector if working conditions are not normal during the inspection.

### Observing monitoring

If health hazards are present in the workplace, a special OSHA health inspection maybe conducted by an industrial hygienist. This OSHA inspector may take samples to measure levels of dust, noise, fumes, or other hazardous materials.

OSHA will inform the employee representative as to whether the employer is in compliance. The inspector also will gather detailed information about the employer's efforts to control health hazards, including results of tests the employer may have conducted.

### *Reviewing OSHA Form 200*

If the employer has more than 10 employees, the employer must maintain records of all work-related injuries and illnesses, and the employees or their representative have the right to review those records. Some industries with very low injury rates (e.g., insurance and real estate offices) are exempt from recordkeeping.

Work-related minor injuries must be recorded if they resulted in restriction of work or motion, loss of consciousness, transfer to another job, termination of employment, or medical treatment (other than first aid). All recognized work-related illnesses and non-minor injuries also must be recorded.

## *After an inspection*

At the end of the inspection, the OSHA inspector will meet with the employer and the employee representatives in a closing conference to discuss the abatement of any hazards that may have been found. If it is not practical to hold a joint conference, separate conferences will be held, and OSHA will provide written summaries, on request.

During the closing conference, the employee representative may describe, if not reported already, what hazards exist, what should be done to correct them, and how long it should take. Other facts about the history of health and safety conditions at the workplace may also be provided.

### *Challenging abatement period*

Whether or not the employer accepts OSHA's actions, the employee (or representative) has the right to contest the time OSHA allows for correcting a hazard. This contest must be filed in writing with the OSHA area director within 15 working days after the citation is issued. The contest will be decided by the Occupational Safety and Health Review Commission. The Review Commission is an independent agency and is not part of the Department of Labor.

### *Variances*

Some employers may not be able to comply fully with a new safety and health standard in the time provided due to shortages of personnel, materials, or equipment. In situations like these, employers may apply to OSHA for a temporary variance from the standard. In other cases, employers may be using methods or equipment that differ from those prescribed by OSHA, but which the employer believes are equal to or better than OSHA's requirements and would qualify for consideration as a permanent variance. Applications for a permanent variance must basically contain the same information as those for temporary variances.

The employer must certify that workers have been informed of the variance application, that a copy has been given to the employee's representative, and that a summary of the application has been posted wherever notices are normally posted in the workplace. Employees also must be informed that they have the right to request a hearing on the application.

Employees, employers, and other interested groups are encouraged to participate in the variance process. Notices of variance application are published in the Federal R*egister,* inviting all interested parties to comment on the action.

### *Confidentiality*

OSHA will not tell the employer who requested the inspection unless the complainant indicates that he or she has no objection.

### *Review if no inspection is made*

The OSHA area director evaluates the complaint from the employee or representative and decides whether it is valid. If the area director decides not to inspect the workplace, he or she will send a certified letter to the complainant explaining the decision and the reasons for it. Complainants must be informed that they have the right to request further clarification of the decision from the area director; if still dissatisfied, they can appeal to the OSHA regional administrator for an informal review. Similarly, a decision by an area director not to issue a citation after an inspection is subject to further clarification from the area director and to an informal review by the regional administrator.

### *Discrimination for using rights*

Employees have a right to seek safety and health on the job without fear of punishment. That right is spelled out in Section 11(c) of the Act. The law says the employer "shall not" punish or discriminate against employees for exercising such rights as complaining to the employer, union, OSHA, or any other government agency about job safety and health hazards; or for participating in OSHA inspections, conferences, hearings, or other OSHA-related activities.

Although there is nothing in the OSHA law that specifically gives an employee the right to refuse to perform an unsafe or unhealthful job assignment, OSHA's regulations, which have been upheld by the U.S. Supreme Court, provide that an employee may refuse to work when faced with an imminent danger of death or serious injury. The conditions necessary to justify a work refusal are very stringent, however, and a work refusal should be an action taken only as a last resort. If time permits, the unhealthful or unsafe condition must be reported to OSHA or other appropriate regulatory agency.

A state that is administering its own occupational safety and health enforcement program pursuant to Section 18 of the Act must have provisions as effective as those of Section 11(c) to protect employees from discharge or discrimination. OSHA, however, retains its Section 11(c) authority in all states regardless of the existence of an OSHA-approved state occupational safety and health program.

Workers believing they have been punished for exercising safety and health rights must contact the nearest OSHA office within 30 days of the time they learn of the alleged discrimination. A representative of the employee's choosing can file the 11(c) complaint for the worker. Following a complaint, OSHA will contact the complainant and conduct an in-depth interview to determine whether an investigation is necessary.

If evidence supports the conclusion that the employee has been punished for exercising safety and health rights, OSHA will ask the employer to restore that worker's job, earnings, and benefits. If the employer declines to enter into a voluntary settlement, OSHA may take the employer to court. In such cases, an attorney of the Department of Labor will conduct litigation on behalf of the employee to obtain this relief.

Section 31105 of the Surface Transportation Assistance Act was enacted on January 6, 1983, and provides protection from reprisal by employers for truckers and certain other employees in the trucking industry involved in activity related to commercial motor vehicle safety and health.

Secretary of Labor's Order No. 9-83 (48 FR 35736, August 5, 1983) delegated to the Assistant Secretary of OSHA the authority to investigate and to issue findings and preliminary orders under Section 31105. Employees who believe they have been discriminated against for exercising their rights under Section 31105 may file a complaint with OSHA within 180 days of the discrimination. OSHA will then investigate the complaint, and within 60 days after it was filed, issue findings as to whether there is a reason to believe Section 31105 has been violated.

If OSHA finds that a complaint has merit, the agency also will issue an order requiring, where appropriate, abatement of the violation, reinstatement with back pay and related compensation, payment of compensatory damages, and payment of the employee's expenses in bringing the complaint. Either the employee or employer may object to the findings. If no objection is filed within 30 days, the finding and order are final. If a timely filed objection is made, however, the objecting party is entitled to a hearing on the objection before an Administrative Law Judge of the Department of Labor. Within 120 days of the hearing, the Secretary will issue a final order. A party aggrieved by the final order may seek judicial review in a court of appeals within 60 days of the final order.

The following activities of truckers and certain employees involved in commercial motor vehicle operation are protected under Section 31105:

- Filing of safety or health complaints with OSHA or other regulatory agency relating to a violation of a commercial motor vehicle safety rule, regulation, standard, or order
- Instituting or causing to be instituted any proceedings relating to a violation of a commercial motor vehicle safety rule, regulation, standard or order
- Testifying in any such proceedings relating to the above items
- Refusing to operate a vehicle when such operation constitutes a violation of any federal rules, regulations, standards, or orders applicable to commercial motor vehicle safety or health; or because of the employee's reasonable apprehension of serious injury to himself or the public due to the unsafe condition of the equipment
- Complaining directly to management, co-workers, or others about job safety or health conditions relating to commercial motor vehicle operation

Complaints under Section 31105 are filed in the same manner as complaints under 11(c). The filing period for Section 31105 is 180 days from the alleged discrimination, rather than 30 days as under Section 11(c).

In addition, Section 211 of the Asbestos Hazard Emergency Response Act provides employee protection from discrimination by school officials in retaliation for complaints about asbestos hazards in primary and secondary schools. The protection and procedures are similar to those used under Section 11(c) of the OSH Act. Section 211 complaints must be filed within 90 days of the alleged discrimination.

Finally, Section 7 of the International Safe Container Act also provides employee protection from discrimination in retaliation for safety or health complaints about intermodal cargo containers designed to be transported interchangeably by sea and land carriers. The protection and procedures are similar to those used under Section 11(c) of the OSH Act. Section 7 complaints must be filed within 60 days of the alleged discrimination.

## *Employee responsibilities*

Although OSHA does not cite employees for violations of their responsibilities, each employee "shall comply with all occupational safety and health standards and all rules, regulations, and orders issued under the Act" that are applicable. Employee responsibilities and rights in states with their own occupational safety and health programs are generally the same as for workers in states covered by federal OSHA. An employee should do the following:

- Read the OSHA poster at the jobsite.
- Comply with all applicable OSHA standards.
- Follow all lawful employer safety and health rules and regulations, and wear or use prescribed protective equipment while working.
- Report hazardous conditions to the supervisor.

- Report any job-related injury or illness to the employer, and seek treatment promptly.
- Cooperate with the OSHA compliance officer conducting an inspection if he or she inquires about safety and health conditions in the workplace.
- Exercise rights under the Act in a responsible manner.

## Contacting NIOSH

NIOSH can provide free information on the potential dangers of substances in the workplace. In some cases, NIOSH may visit a jobsite to evaluate possible health hazards. The address is as follows:

National Institute for Occupational Safety and Health
Centers for Disease Control
1600 Clifton Road
Atlanta, GA 30333
Telephone: 404-639-3061

NIOSH will keep confidential the name of the person who asked for help if requested to do so.

## Safety and health management guidelines

Effective management of worker safety and health protection is a decisive factor in reducing the extent and severity of work-related injuries and illnesses and their related costs. To assist employers and employees in developing effective safety and health programs, OSHA has published recommended safety and health program management guidelines (*Federal Register* 54(16): 3904–3916, January 26, 1989). These voluntary guidelines apply to all places of employment covered by OSHA.

The guidelines identify four general elements that are critical to the development of a successful safety and health management program:

- Management commitment and employee involvement
- Worksite analysis
- Hazard prevention and control
- Safety and health training

The guidelines recommend specific actions, under each of these general elements, to achieve an effective safety and health program. A single free copy of the guidelines can be obtained from the OSHA Publications Office, U.S. Department of Labor, OSHA/OSHA Publications, P.O. Box 37535, Washington, D.C. 20013-7535, by sending a self-addressed mail label with your request.

# *appendix C*

# *Respiratory equipment**

## *Appendix A to §1910.134: fit testing procedures (mandatory)*

### *Part I. OSHA-accepted fit test protocols*

A. *Fit testing procedures — general requirements:* The employer shall conduct fit testing using the following procedures. The requirements in this appendix apply to all OSHA-accepted fit test methods, both qualitative fit test (QLFT) and quantitative fit test (QNFT).

1. The test subject shall be allowed to pick the most acceptable respirator from a sufficient number of respirator models and sizes so that the respirator is acceptable to, and correctly fits, the user.
2. Prior to the selection process, the test subject shall be shown how to put on a respirator, how it should be positioned on the face, how to set strap tension, and how to determine an acceptable fit. A mirror shall be available to assist the subject in evaluating the fit and positioning of the respirator. This instruction may not constitute the subject's formal training on respirator use, because it is only a review.
3. The test subject shall be informed that he/she is being asked to select the respirator that provides the most acceptable fit. Each respirator represents a different size and shape and, if fitted and used properly, will provide adequate protection.
4. The test subject shall be instructed to hold each chosen facepiece up to the face and eliminate those that obviously do not give an acceptable fit.

* From http://www.osha_slc.gov/oshstd_data/1910_0134_app_a.html.

5. The more acceptable facepieces are noted in case the one selected proves unacceptable; the most comfortable mask is donned and worn at least 5 minutes to assess comfort. Assistance in assessing comfort can be given by discussing the points in the following item (A.6). If the test subject is not familiar with using a particular respirator, the test subject shall be directed to don the mask several times and to adjust the straps each time to become adept at setting proper tension on the straps.

6. Assessment of comfort shall include a review of the following points with the test subject and allowing the test subject adequate time to determine the comfort of the respirator:

   (a) Position of the mask on the nose

   (b) Room for eye protection

   (c) Room to talk

   (d) Position of mask on face and cheeks

7. The following criteria shall be used to help determine the adequacy of the respirator fit:

   (a) Chin properly placed

   (b) Adequate strap tension, not overly tightened

   (c) Fit across nose bridge

   (d) Respirator of proper size to span distance from nose to chin

   (e) Tendency of respirator to slip

   (f) Self-observation in mirror to evaluate fit and respirator position

8. The test subject shall conduct a user seal check, either the negative and positive pressure seal checks described in Appendix B-1 of this section or those recommended by the respirator manufacturer which provide equivalent protection to the procedures in Appendix B-1. Before conducting the negative and positive pressure checks, the subject shall be told to seat the mask on the face by moving the head from side-to-side and up and down slowly while taking in a few slow deep breaths. Another facepiece shall be selected and retested if the test subject fails the user seal check tests.

9. The test shall not be conducted if there is any hair growth between the skin and the facepiece sealing surface, such as stubble beard growth, beard, mustache, or sideburns which cross the respirator sealing surface. Any type of apparel that interferes with a satisfactory fit shall be altered or removed.

10. If a test subject exhibits difficulty in breathing during the tests, she or he shall be referred to a physician or other licensed health-care professional, as appropriate, to determine whether the test subject can wear a respirator while performing his or her duties.

11. If the employee finds the fit of the respirator unacceptable, the test subject shall be given the opportunity to select a different respirator and to be retested.

12. *Exercise regimen:* Prior to commencement of the fit test, the test subject shall be given a description of the fit test and the test subject's responsibilities during the test procedure. The description of the process shall include a description of the test exercises that the subject will be performing. The respirator to be tested shall be worn for at least 5 minutes before the start of the fit test.

13. The fit test shall be performed while the test subject is wearing any applicable safety equipment that may be worn during actual respirator use which could interfere with respirator fit.

14. *Test exercises:*

(a) The following test exercises are to be performed for all fit testing methods prescribed in this appendix, except for the CNP method. A separate fit testing exercise regimen is contained in the CNP protocol. The test subject shall perform exercises, in the test environment, in the following manner:

(1) *Normal breathing:* In a normal standing position, without talking, the subject shall breathe normally.

(2) *Deep breathing:* In a normal standing position, the subject shall breathe slowly and deeply, taking caution so as not to hyperventilate.

(3) *Turning head side to side:* Standing in place, the subject shall slowly turn his/her head from side to side between the extreme positions on each side. The head shall be held at each extreme momentarily so the subject can inhale at each side.

(4) *Moving head up and down:* Standing in place, the subject shall slowly move his or her head up and down. The subject shall be instructed to inhale in the up position (i.e., when looking toward the ceiling).

(5) *Talking:* The subject shall talk out loud slowly and loud enough so as to be heard clearly by the test conductor. The subject can read from a prepared text such as the "Rainbow Passage," count backward from 100, or recite a memorized poem or song.

*Rainbow Passage*
When the sunlight strikes raindrops in the air, they act like a prism and form a rainbow. The rainbow is a division of white light into many beautiful colors. These take the shape of a long round arch, with its path high above, and its two ends apparently beyond the horizon. There is, according

to legend, a boiling pot of gold at one end. People look, but no one ever finds it. When a man looks for something beyond reach, his friends say he is looking for the pot of gold at the end of the rainbow.

(6) *Grimace:* The test subject shall grimace by smiling or frowning. (This applies only to QNFT testing; it is not performed for QLFT)

(7) *Bending over:* The test subject shall bend at the waist as if he/she were to touch his/her toes. Jogging in place shall be substituted for this exercise in those test environments such as shroud-type QNFT or QLFT units that do not permit bending over at the waist.

(8) *Normal breathing:* Same as exercise (1).

(b) Each test exercise shall be performed for 1 minute, except for the grimace exercise, which shall be performed for 15 seconds. The test subject shall be questioned by the test conductor regarding the comfort of the respirator upon completion of the protocol. If it has become unacceptable, another model of respirator shall be tried. The respirator shall not be adjusted once the fit test exercises begin. Any adjustment voids the test, and the fit test must be repeated.

B. *Qualitative fit test (QLFT) protocols:*

1. *General:*

(a) The employer shall ensure that persons administering QLFT are able to prepare test solutions, calibrate equipment, perform tests properly, recognize invalid tests, and ensure that test equipment is in proper working order.

(b) The employer shall ensure that QLFT equipment is kept clean and well maintained so as to operate within the parameters for which it was designed.

2. *Isoamyl acetate protocol: (Note:* This protocol is not appropriate to use for the fit testing of particulate respirators. If used to fit test particulate respirators, the respirator must be equipped with an organic vapor filter.)

(a) *Odor threshold screening:* Odor threshold screening, performed without wearing a respirator, is intended to determine if the individual tested can detect the odor of isoamyl acetate at low levels.

(1) Three 1-liter glass jars with metal lids are required.

(2) Odor-free water (e.g., distilled or spring water) at approximately 25 deg. C (77 deg. F) shall be used for the solutions.

(3) The isoamyl acetate (IAA) (also known at isopentyl acetate) stock solution is prepared by adding 1 ml of pure IAA to 800 ml of odor-free water in a 1-liter jar, closing the lid, and shaking for 30 seconds. A new solution shall be prepared at least weekly.

(4) The screening test shall be conducted in a room separate from the room used for actual fit testing. The two rooms shall be well-ventilated to prevent the odor of IAA from becoming evident in the general room air where testing takes place.

(5) The odor test solution is prepared in a second jar by placing 0.4 ml of the stock solution into 500 ml of odor-free water using a clean dropper or pipette. The solution shall be shaken for 30 seconds and allowed to stand for 2 to 3 minutes so that the IAA concentration above the liquid may reach equilibrium. This solution shall be used for only one day.

(6) A test blank shall be prepared in a third jar by adding 500 cc of odor-free water.

(7) The odor test and test blank jar lids shall be labeled (e.g., 1 and 2) for jar identification. Labels shall be placed on the lids so that they can be peeled off periodically and switched to maintain the integrity of the test.

(8) The following instruction shall be typed on a card and placed on the table in front of the two test jars (i.e., 1 and 2): "The purpose of this test is to determine if you can smell banana oil at a low concentration. The two bottles in front of you contain water. One of these bottles also contains a small amount of banana oil. Be sure the covers are on tight, then shake each bottle for two seconds. Unscrew the lid of each bottle, one at a time, and sniff at the mouth of the bottle. Indicate to the test conductor which bottle contains banana oil."

(9) The mixtures used in the IAA odor detection test shall be prepared in an area separate from where the test is performed, in order to prevent olfactory fatigue in the subject.

(10) If the test subject is unable to correctly identify the jar containing the odor test solution, the IAA qualitative fit test shall not be performed.

(11) If the test subject correctly identifies the jar containing the odor test solution, the test subject may proceed to respirator selection and fit testing.

(b) *Isoamyl acetate fit test:*

(1) The fit test chamber shall be a clear 55-gallon drum liner suspended inverted over a 2-foot-diameter frame so that the top of the chamber is about 6 inches above the test subject's head. If no drum liner is available, a similar chamber shall be constructed using plastic sheeting. The inside top center of the chamber shall have a small hook attached.

(2) Each respirator used for the fitting and fit testing shall be equipped with organic vapor cartridges or offer protection against organic vapors. Check the respirator to make sure the sampling probe and line are properly attached to the facepiece and that the respirator is fitted with a particulate filter capable of preventing significant penetration by the ambient particles used for the fit test (e.g., NIOSH 42 C.F.R. 84 series 100, series 99, or series 95 particulate filter) per manufacturer's instruction. Instruct the person to be tested to don the respirator for 5 minutes before the fit test starts. This purges the ambient particlead. If no drum liner is available, a similar chamber shall be constructed using plastic sheeting. The inside top center of the chamber shall have a small hook attached.

(3) After selecting, donning, and properly adjusting a respirator, the test subject shall wear it to the fit testing room. This room shall be separate from the room used for odor threshold screening and respirator selection and shall be well-ventilated, as by an exhaust fan or lab hood, to prevent general room contamination.

(4) A copy of the test exercises and any prepared text from which the subject is to read shall be taped to the inside of the test chamber.

(5) Upon entering the test chamber, the test subject shall be given a 6-inch by 5-inch piece of paper towel or other porous, absorbent, single-ply material, folded in half and wetted with 0.75 ml of pure IAA. The test subject shall hang the wet towel on the hook at the top of the chamber. An IAA test swab or ampule may be substituted for the IAA wetted paper towel provided it has been demonstrated that the alternative IAA source will generate an IAA test atmosphere with a concentration equivalent to that generated by the paper towel method.

(6) Allow 2 minutes for the IAA test concentration to stabilize before starting the fit test exercises. This would be an appropriate time to talk with the test subject; to explain

the fit test, the importance of his/her cooperation, and the purpose for the test exercises; or to demonstrate some of the exercises.

(7) If, at any time during the test, the subject detects the banana-like odor of IAA, the test is failed. The subject shall quickly exit from the test chamber and leave the test area to avoid olfactory fatigue.

(8) If the test is failed, the subject shall return to the selection room and remove the respirator. The test subject shall repeat the odor sensitivity test, select and put on another respirator, return to the test area, and again begin the fit test procedure described in (b)(1) through (7) above. The process continues until a respirator that fits well has been found. Should the odor sensitivity test be failed, the subject shall wait at least 5 minutes before retesting. Odor sensitivity will usually have returned by this time.

(9) If the subject passes the test, the efficiency of the test procedure shall be demonstrated by having the subject break the respirator face seal and take a breath before exiting the chamber.

(10) When the test subject leaves the chamber, the subject shall remove the saturated towel and return it to the person conducting the test so that there is no significant IAA concentration build-up in the chamber during subsequent tests. The used towels shall be kept in a self-sealing plastic bag to keep the test area from being contaminated.

3. *Saccharin solution aerosol protocol:* The entire screening and testing procedure shall be explained to the test subject prior to conducting the screening test.

(a) *Taste threshold screening:* The saccharin taste threshold screening, performed without wearing a respirator, is intended to determine whether the individual being tested can detect the taste of saccharin.

(1) During threshold screening, as well as during fit testing, subjects shall wear an enclosure about the head and shoulders that is approximately 12 inches in diameter by 14 inches tall with at least the front portion clear and that allows free movements of the head when a respirator is worn. An enclosure substantially similar to the 3M hood assembly, parts #FT 14 and #FT 15 combined, is adequate.

(2) The test enclosure shall have a 3/4-inch (1.9-cm) hole in front of the test subject's nose and mouth area to accommodate the nebulizer nozzle.

(3) The test subject shall don the test enclosure. Throughout the threshold screening test, the test subject shall breathe through his/her slightly open mouth with tongue extended. The subject is instructed to report when he/she detects a sweet taste.

(4) Using a DeVilbiss Model 40 Inhalation Medication Nebulizer or equivalent, the test conductor shall spray the threshold check solution into the enclosure. The nozzle is directed away from the nose and mouth of the person. This nebulizer shall be clearly marked to distinguish it from the fit test solution nebulizer.

(5) The threshold check solution is prepared by dissolving 0.83 gram of sodium saccharin USP in 100 ml of warm water. It can be prepared by putting 1 ml of the fit test solution (see (b)(5) below) in 100 ml of distilled water.

(6) To produce the aerosol, the nebulizer bulb is firmly squeezed so that it collapses completely, then released and allowed to fully expand.

(7) Ten squeezes are repeated rapidly and then the test subject is asked whether the saccharin can be tasted. If the test subject reports tasting the sweet taste during the ten squeezes, the screening test is completed. The taste threshold is noted as ten regardless of the number of squeezes actually completed.

(8) If the first response is negative, ten more squeezes are repeated rapidly and the test subject is again asked whether the saccharin is tasted. If the test subject reports tasting the sweet taste during the second ten squeezes, the screening test is completed. The taste threshold is noted as twenty regardless of the number of squeezes actually completed.

(9) If the second response is negative, ten more squeezes are repeated rapidly and the test subject is again asked whether the saccharin is tasted. If the test subject reports tasting the sweet taste during the third set of ten squeezes, the screening test is completed. The taste threshold is noted as thirty regardless of the number of squeezes actually completed.

(10) The test conductor will take note of the number of squeezes required to solicit a taste response.

(11) If the saccharin is not tasted after 30 squeezes (step 10), the test subject is unable to taste saccharin and may not perform the saccharin fit test.

*Note to paragraph 3.(a):* If the test subject eats or drinks something sweet before the screening test, he/she may be unable to taste the weak saccharin solution.

(12) If a taste response is elicited, the test subject shall be asked to take note of the taste for reference in the fit test.

(13) Correct use of the nebulizer means that approximately 1 ml of liquid is used at a time in the nebulizer body.

(14) The nebulizer shall be thoroughly rinsed in water, shaken dry, and refilled at least each morning and afternoon or at least every four hours.

(b) *Saccharin solution aerosol fit test procedure:*

(1) The test subject may not eat, drink (except plain water), smoke, or chew gum for 15 minutes before the test.

(2) The fit test uses the same enclosure described in 3.(a) above.

(3) The test subject shall don the enclosure while wearing the respirator selected in Section I.A of this appendix. The respirator shall be properly adjusted and equipped with a particulate filter(s).

(4) A second DeVilbiss Model 40 Inhalation Medication Nebulizer or equivalent is used to spray the fit test solution into the enclosure. This nebulizer shall be clearly marked to distinguish it from the screening test solution nebulizer.

(5) The fit test solution is prepared by adding .83 grams of sodium saccharin to 100 ml of warm water.

(6) As before, the test subject shall breathe through the slightly open mouth with tongue extended and report if he/she tastes the sweet taste of saccharin.

(7) The nebulizer is inserted into the hole in the front of the enclosure and an initial concentration of saccharin fit test solution is sprayed into the enclosure using the same number of squeezes (10, 20, or 30 squeezes) based on the number of squeezes required to elicit a taste response as noted during the screening test. A minimum of 10 squeezes is required.

(8) After generating the aerosol, the test subject shall be instructed to perform the exercises in Section I.A.14 of this appendix.

(9) Every 30 seconds the aerosol concentration shall be replenished using one half the original number of squeezes used initially (e.g., 5, 10, or 15).

(10) The test subject shall indicate to the test conductor if at any time during the fit test the taste of saccharin is detected. If the test subject does not report tasting the saccharin, the test is passed.

(11) If the taste of saccharin is detected, the fit is deemed unsatisfactory and the test is failed. A different respirator shall be tried and the entire test procedure is repeated (taste threshold screening and fit testing).

(12) Since the nebulizer has a tendency to clog during use, the test operator must make periodic checks of the nebulizer to ensure that it is not clogged. If clogging is found at the end of the test session, the test is invalid.

4. *Bitrex™ (denatonium benzoate) solution aerosol qualitative fit test protocol:* The Bitrex™ (denatonium benzoate) solution aerosol QLFT protocol uses the published saccharin test protocol because that protocol is widely accepted. Bitrex™ is routinely used as a taste-aversion agent in household liquids that children should not be drinking and is endorsed by the American Medical Association, the National Safety Council, and the American Association of Poison Control Centers. The entire screening and testing procedure shall be explained to the test subject prior to conducting the screening test.

(a) *Taste threshold screening:* The Bitrex™ taste threshold screening, performed without wearing a respirator, is intended to determine whether the individual being tested can detect the taste of Bitrex™.

(1) During threshold screening as well as during fit testing, subjects shall wear an enclosure about the head and shoulders that is approximately 12 inches (30.5 cm) in diameter by 14 inches (35.6 cm) tall. The front portion of the enclosure shall be clear from the respirator and allow free movement of the head when a respirator is worn. An enclosure substantially similar to the 3M hood assembly, parts #FT 14 and #FT 15 combined, is adequate.

(2) The test enclosure shall have a 3/4-inch (1.9-cm) hole in front of the test subject's nose and mouth area to accommodate the nebulizer nozzle.

(3) The test subject shall don the test enclosure. Throughout the threshold screening test, the test subject shall breathe through his or her slightly open mouth with tongue extended. The subject is instructed to report when he/she detects a bitter taste

(4) Using a DeVilbiss Model 40 Inhalation Medication Nebulizer or equivalent, the test conductor shall spray the

Threshold Check Solution into the enclosure. This Nebulizer shall be clearly marked to distinguish it from the fit test solution nebulizer.

(5) The Threshold Check Solution is prepared by adding 13.5 mg of Bitrex™ to 100 ml of 5% salt (NaCl) solution in distilled water.

(6) To produce the aerosol, the nebulizer bulb is firmly squeezed, so that the bulb collapses completely, and is then released and allowed to fully expand.

(7) An initial ten squeezes are repeated rapidly and then the test subject is asked whether the Bitrex™ can be tasted. If the test subject reports tasting the bitter taste during the ten squeezes, the screening test is completed. The taste threshold is noted as ten regardless of the number of squeezes actually completed.

(8) If the first response is negative, ten more squeezes are repeated rapidly and the test subject is again asked whether the Bitrex™ is tasted. If the test subject reports tasting the bitter taste during the second ten squeezes, the screening test is completed. The taste threshold is noted as twenty regardless of the number of squeezes actually completed.

(9) If the second response is negative, ten more squeezes are repeated rapidly and the test subject is again asked whether the Bitrex™ is tasted. If the test subject reports tasting the bitter taste during the third set of ten squeezes, the screening test is completed. The taste threshold is noted as thirty regardless of the number of squeezes actually completed.

(10) The test conductor will take note of the number of squeezes required to solicit a taste response.

(11) If the Bitrex™ is not tasted after 30 squeezes (step 10), the test subject is unable to taste Bitrex™ and may not perform the Bitrex™ fit test.

(12) If a taste response is elicited, the test subject shall be asked to take note of the taste for reference in the fit test.

(13) Correct use of the nebulizer means that approximately 1 ml of liquid is used at a time in the nebulizer body.

(14) The nebulizer shall be thoroughly rinsed in water, shaken to dry, and refilled at least each morning and afternoon or at least every four hours.

(b) *Bitrex™ solution aerosol fit test procedure:*

(1) The test subject may not eat, drink (except plain water), smoke, or chew gum for 15 minutes before the test.

(2) The fit test uses the same enclosure as that described in 4.(a) above.

(3) The test subject shall don the enclosure while wearing the respirator selected according to Section I.A of this appendix. The respirator shall be properly adjusted and equipped with any type particulate filter(s).

(4) A second DeVilbiss Model 40 Inhalation Medication Nebulizer or equivalent is used to spray the fit test solution into the enclosure. This nebulizer shall be clearly marked to distinguish it from the screening test solution nebulizer.

(5) The fit test solution is prepared by adding 337.5 mg of Bitrex™ to 200 ml of a 5% salt (NaCl) solution in warm water.

(6) As before, the test subject shall breathe through his or her slightly open mouth with tongue extended and be instructed to report if he/she tastes the bitter taste of Bitrex™.

(7) The nebulizer is inserted into the hole in the front of the enclosure and an initial concentration of the fit test solution is sprayed into the enclosure using the same number of squeezes (10, 20, or 30 squeezes) based on the number of squeezes required to elicit a taste response as noted during the screening test.

(8) After generating the aerosol, the test subject shall be instructed to perform the exercises in Section I.A.14 of this appendix.

(9) Every 30 seconds the aerosol concentration shall be replenished using one half the number of squeezes used initially (e.g., 5, 10, or 15).

(10) The test subject shall indicate to the test conductor if at any time during the fit test the taste of Bitrex™ is detected. If the test subject does not report tasting the Bitrex™, the test is passed.

(11) If the taste of Bitrex™ is detected, the fit is deemed unsatisfactory and the test is failed. A different respirator shall be tried and the entire test procedure is repeated (taste threshold screening and fit testing).

5. *Irritant smoke (stannic chloride) protocol:* This qualitative fit test uses a person's response to the irritating chemicals released in the "smoke" produced by a stannic chloride ventilation smoke tube to detect leakage into the respirator.

(a) *General requirements and precautions:*

(1) The respirator to be tested shall be equipped with high-efficiency particulate air (HEPA) or P100 series filter(s).

(2) Only stannic chloride smoke tubes shall be used for this protocol.

(3) No form of test enclosure or hood for the test subject shall be used.

(4) The smoke can be irritating to the eyes, lungs, and nasal passages. The test conductor shall take precautions to minimize the test subject's exposure to irritant smoke. Sensitivity varies, and certain individuals may respond to a greater degree to irritant smoke. Care shall be taken when performing the sensitivity screening checks that determine whether the test subject can detect irritant smoke to use only the minimum amount of smoke necessary to elicit a response from the test subject.

(5) The fit test shall be performed in an area with adequate ventilation to prevent exposure of the person conducting the fit test or the build-up of irritant smoke in the general atmosphere.

(b) *Sensitivity screening check:* The person to be tested must demonstrate his or her ability to detect a weak concentration of the irritant smoke.

(1) The test operator shall break both ends of a ventilation smoke tube containing stannic chloride and attach one end of the smoke tube to a low-flow air pump set to deliver 200 ml per minute or to an aspirator squeeze bulb. The test operator shall cover the other end of the smoke tube with a short piece of tubing to prevent potential injury from the jagged end of the smoke tube.

(2) The test operator shall advise the test subject that the smoke can be irritating to the eyes, lungs, and nasal passages and instruct the subject to keep his/her eyes closed while the test is performed.

(3) The test subject shall be allowed to smell a weak concentration of the irritant smoke before the respirator is donned to become familiar with its irritating properties and to determine if he/she can detect the irritating properties of the smoke. The test operator shall carefully direct a small amount of the irritant smoke in the test subject's direction to determine that he/she can detect it.

(c) *Irritant smoke fit test procedure:*

(1) The person being fit tested shall don the respirator without assistance and perform the required user seal check(s).

(2) The test subject shall be instructed to keep his/her eyes closed.

(3) The test operator shall direct the stream of irritant smoke from the smoke tube toward the faceseal area of the test subject, using the low flow pump or the squeeze bulb. The test operator shall begin at least 12 inches from the facepiece and move the smoke stream around the whole perimeter of the mask. The operator shall gradually make two more passes around the perimeter of the mask, moving to within 6 inches of the respirator.

(4) If the person being tested has not had an involuntary response and/or detected the irritant smoke, proceed with the test exercises.

(5) The exercises identified in Section I.A.14 of this appendix shall be performed by the test subject while the respirator seal is being continually challenged by the smoke, directed around the perimeter of the respirator at a distance of 6 inches.

(6) If the person being fit tested reports detecting the irritant smoke at any time, the test is failed. The person being retested must repeat the entire sensitivity check and fit test procedure.

(7) Each test subject passing the irritant smoke test without evidence of a response (involuntary cough, irritation) shall be given a second sensitivity screening check, with the smoke from the same smoke tube used during the fit test, once the respirator has been removed, to determine whether he/she still reacts to the smoke. Failure to evoke a response shall void the fit test.

(8) If a response is produced during this second sensitivity check, then the fit test is passed.

C. *Quantitative fit test (QNFT) protocols:* The following quantitative fit testing procedures have been demonstrated to be acceptable: Quantitative fit testing using a non-hazardous test aerosol (such as corn oil, polyethylene glycol 400 [PEG 400], di-2-ethyl hexyl sebacate [DEHS], or sodium chloride) generated in a test chamber, and employing instrumentation to quantify the fit of the respirator; quantitative fit testing using ambient aerosol as the test agent and appropriate instrumentation (condensation nuclei counter) to quantify the respirator fit; quantitative fit testing using controlled negative pressure and appropriate instrumentation to measure the volumetric leak rate of a facepiece to quantify the respirator fit.

1. *General:*

(a) The employer shall ensure that persons administering QNFT are able to calibrate equipment and perform tests properly, recognize invalid tests, calculate fit factors properly, and ensure that test equipment is in proper working order.

(b) The employer shall ensure that QNFT equipment is kept clean and is maintained and calibrated according to the manufacturer's instructions so as to operate at the parameters for which it was designed.

2. Generated aerosol quantitative fit testing protocol

(a) *Apparatus:*

(1) *Instrumentation:* Aerosol generation, dilution, and measurement systems using particulates (corn oil, polyethylene glycol 400 [PEG 400], di-2-ethyl hexyl sebacate [DEHS], or sodium chloride) as test aerosols shall be used for quantitative fit testing.

(2) *Test chamber:* The test chamber shall be large enough to permit all test subjects to perform freely all required exercises without disturbing the test agent concentration or the measurement apparatus. The test chamber shall be equipped and constructed so that the test agent is effectively isolated from the ambient air, yet uniform in concentration throughout the chamber.

(3) When testing air-purifying respirators, the normal filter or cartridge element shall be replaced with a high-efficiency particulate air (HEPA) or P100 series filter supplied by the same manufacturer.

(4) The sampling instrument shall be selected so that a computer record or strip chart record may be made of the test showing the rise and fall of the test agent concentration with each inspiration and expiration at fit factors of at least 2000. Integrators or computers that integrate the amount of test agent penetration leakage into the respirator for each exercise may be used provided a record of the readings is made.

(5) The combination of substitute air-purifying elements, test agent, and test agent concentration shall be such that the test subject is not exposed in excess of an established exposure limit for the test agent at any time during the testing process, based upon the length of the exposure and the exposure limit duration.

(6) The sampling port on the test specimen respirator shall be placed and constructed so that no leakage occurs around the port (e.g., where the respirator is probed), a free air flow is allowed into the sampling line at all times, and there is no interference with the fit or performance of the respirator. The in-mask sampling device (probe) shall be

designed and used so that the air sample is drawn from the breathing zone of the test subject, midway between the nose and mouth and with the probe extending into the facepiece cavity at least 1/4 inch.

(7) The test setup shall permit the person administering the test to observe the test subject inside the chamber during the test.

(8) The equipment generating the test atmosphere shall maintain the concentration of test agent constant to within a 10 percent variation for the duration of the test.

(9) The time lag (interval between an event and the recording of the event on the strip chart or computer or integrator) shall be kept to a minimum. There shall be a clear association between the occurrence of an event and its being recorded.

(10) The sampling line tubing for the test chamber atmosphere and for the respirator sampling port shall be of equal diameter and of the same material. The length of the two lines shall be equal.

(11) The exhaust flow from the test chamber shall pass through an appropriate filter (i.e., high-efficiency particulate filter) before release.

(12) When sodium chloride aerosol is used, the relative humidity inside the test chamber shall not exceed 50 percent.

(13) The limitations of instrument detection shall be taken into account when determining the fit factor.

(14) Test respirators shall be maintained in proper working order and be inspected regularly for deficiencies such as cracks or missing valves and gaskets.

(b) *Procedural requirements:*

(1) When performing the initial user seal check using a positive or negative pressure check, the sampling line shall be crimped closed in order to avoid air pressure leakage during either of these pressure checks.

(2) The use of an abbreviated screening QLFT test is optional. Such a test may be utilized in order to quickly identify poor-fitting respirators that passed the positive and/or negative pressure test and reduce the amount of QNFT time. The use of the CNC QNFT instrument in the count mode is another optional method to obtain a quick estimate of fit and eliminate poor-fitting respirators before going on to perform a full QNFT.

(3) A reasonably stable test agent concentration shall be measured in the test chamber prior to testing. For canopy or shower curtain types of test units, the determination of the test agent's stability may be established after the test subject has entered the test environment.

(4) Immediately after the subject enters the test chamber, the test agent concentration inside the respirator shall be measured to ensure that the peak penetration does not exceed 5 percent for a half mask or 1 percent for a full facepiece respirator.

(5) A stable test agent concentration shall be obtained prior to the actual start of testing.

(6) Respirator restraining straps shall not be over-tightened for testing. The straps shall be adjusted by the wearer without assistance from other persons to give a reasonably comfortable fit typical of normal use. The respirator shall not be adjusted once the fit test exercises begin.

(7) The test shall be terminated whenever any single peak penetration exceeds 5 percent for half masks and 1 percent for full facepiece respirators. The test subject shall be refitted and retested.

(8) *Calculation of fit factors:*

(i) The fit factor shall be determined for the quantitative fit test by taking the ratio of the average chamber concentration to the concentration measured inside the respirator for each test exercise except the grimace exercise.

(ii) The average test chamber concentration shall be calculated as the arithmetic average of the concentration measured before and after each test (i.e., 7 exercises) or the arithmetic average of the concentration measured before and after each exercise or the true average measured continuously during the respirator sample.

(iii) The concentration of the challenge agent inside the respirator shall be determined by one of the following methods:

*(A)* Average peak penetration method means the method of determining test agent penetration into the respirator utilizing a strip chart recorder, integrator, or computer. The agent penetration is determined by an average of the peak heights on the graph or by computer integration, for each exercise

except the grimace exercise. Integrators or computers that calculate the actual test agent penetration into the respirator for each exercise will also be considered to meet the requirements of the average peak penetration method.

*(B)* Maximum peak penetration method means the method of determining test agent penetration in the respirator as determined by strip chart recordings of the test. The highest peak penetration for a given exercise is taken to be representative of average penetration into the respirator for that exercise.

*(C)* Integration by calculation of the area under the individual peak for each exercise except the grimace exercise includes computerized integration.

*(D)* The calculation of the overall fit factor using individual exercise fit factors involves first converting the exercise fit factors to penetration values, determining the average, and then converting that result back to a fit factor. This procedure is described in the following equation:

$$\text{Overall Fit Factor} = \frac{\text{Number of exercises}}{1/ff_1 + 1/ff_2 + 1/ff_3 + 1/ff_4 + 1/ff_5 + 1/ff_6 + 1/ff_7 + 1/ff_8}$$

Where $ff_1$, $ff_2$, $ff_3$, etc. are the fit factors for exercises 1, 2, 3, etc.

(9) The test subject shall not be permitted to wear a half mask or quarter facepiece respirator unless a minimum fit factor of 100 is obtained, or a full facepiece respirator unless a minimum fit factor of 500 is obtained.

(10) Filters used for quantitative fit testing shall be replaced whenever increased breathing resistance is encountered, or when the test agent has altered the integrity of the filter media.

3. *Ambient aerosol condensation nuclei counter (CNC) quantitative fit testing protocol:* The ambient aerosol condensation nuclei counter (CNC) quantitative fit testing (Portacount™) protocol quantitatively fit tests respirators with the use of a probe. The probed respirator is only used for quantitative fit tests. A probed respirator has a special sampling device, installed on the respirator, that allows the probe to sample the air from inside the mask. A probed respirator is required for each make, style, model, and size that the employer uses and can be obtained from the respirator manufacturer or distributor. The CNC instrument manufacturer, TSI Inc., also provides

probe attachments (TSI sampling adapters) that permit fit testing in an employee's own respirator. A minimum fit factor pass level of at least 100 is necessary for a half-mask respirator, and a minimum fit factor pass level of at least 500 is required for a full facepiece negative pressure respirator. The entire screening and testing procedure shall be explained to the test subject prior to conducting the screening test.

(a) *Portacount™ fit test requirements:*

(1) Check the respirator to make sure the sampling probe and line are properly attached to the facepiece and that the respirator is fitted with a particulate filter capable of preventing significant penetration by the ambient particles used for the fit test (e.g., NIOSH 42 C.F.R. 84 series 100, series 99, or series 95 particulate filter) per manufacturer's instructions.

(2) Instruct the person to be tested to don the respirator for 5 minutes before the fit test starts. This purges the ambient particles trapped inside the respirator and permits the wearer to make certain the respirator is comfortable. This individual shall already have been trained on how to wear the respirator properly.

(3) Check the following conditions for the adequacy of the respirator fit: Chin properly placed; adequate strap tension, not overly tightened; fit across nose bridge; respirator of proper size to span distance from nose to chin; tendency of the respirator to slip; self-observation in a mirror to evaluate fit and respirator position.

(4) Have the person wearing the respirator do a user seal check. If leakage is detected, determine the cause. If leakage is from a poorly fitting facepiece, try another size of the same model respirator, or another model of respirator.

(5) Follow the manufacturer's instructions for operating the Portacount™ and proceed with the test.

(6) The test subject shall be instructed to perform the exercises in Section I.A.14 of this appendix.

(7) After the test exercises, the test subject shall be questioned by the test conductor regarding the comfort of the respirator upon completion of the protocol. If it has become unacceptable, another model of respirator shall be tried.

(b) *Portacount™ test instrument:*

(1) The Portacount™ will automatically stop and calculate the overall fit factor for the entire set of exercises. The overall fit factor is what counts. The Pass or Fail message

will indicate whether or not the test was successful. If the test was a Pass, the fit test is over.

(2) Since the pass or fail criterion of the Portacount™ is user programmable, the test operator shall ensure that the pass or fail criteria meet the requirements for minimum respirator performance in this appendix.

(3) A record of the test needs to be kept on file, assuming the fit test was successful. The record must contain the test subject's name; overall fit factor; make, model, style, and size of respirator used; and date tested.

4. *Controlled negative pressure (CNP) quantitative fit testing protocol:* The CNP protocol provides an alternative to aerosol fit test methods. The CNP fit test method technology is based on exhausting air from a temporarily sealed respirator facepiece to generate and then maintain a constant negative pressure inside the facepiece. The rate of air exhaust is controlled so that a constant negative pressure is maintained in the respirator during the fit test. The level of pressure is selected to replicate the mean inspiratory pressure that causes leakage into the respirator under normal use conditions. With pressure held constant, air flow out of the respirator is equal to air flow into the respirator. Therefore, measurement of the exhaust stream that is required to hold the pressure in the temporarily sealed respirator constant yields a direct measure of leakage air flow into the respirator. The CNP fit test method measures leak rates through the facepiece as a method for determining the facepiece fit for negative pressure respirators. The CNP instrument manufacturer, Dynatech Nevada, also provides attachments (sampling manifolds) that replace the filter cartridges to permit fit testing in an employee's own respirator. To perform the test, the test subject closes his/her mouth and holds his/her breath, after which an air pump removes air from the respirator facepiece at a pre-selected constant pressure. The facepiece fit is expressed as the leak rate through the facepiece, expressed as milliliters per minute. The quality and validity of the CNP fit tests are determined by the degree to which the in-mask pressure tracks the test pressure during the system measurement time of approximately five seconds. Instantaneous feedback in the form of a real-time pressure trace of the in-mask pressure is provided and used to determine test validity and quality. A minimum fit factor pass level of 100 is necessary for a half-mask respirator and a minimum fit factor of at least 500 is required for a full facepiece respirator. The entire screening and testing procedure shall be explained to the test subject prior to conducting the screening test.

(a) *CNP fit test requirements:*

(1) The instrument shall have a non-adjustable test pressure of 15.0 mm water pressure.

(2) The CNP system defaults selected for test pressure shall be set at –15 mm of water (–0.58 inches of water) and the modeled inspiratory flow rate shall be 53.8 liters per minute for performing fit tests. (*Note:* CNP systems have built-in capability to conduct fit testing that is specific to unique work rate, mask, and gender situations that might apply in a specific workplace. Use of system default values, which were selected to represent respirator wear with medium cartridge resistance at a low-moderate work rate, will allow inter-test comparison of the respirator fit.)

(3) The individual who conducts the CNP fit testing shall be thoroughly trained to perform the test.

(4) The respirator filter or cartridge needs to be replaced with the CNP test manifold. The inhalation valve downstream from the manifold either needs to be temporarily removed or propped open.

(5) The test subject shall be trained to hold his or her breath for at least 20 seconds.

(6) The test subject shall don the test respirator without any assistance from the individual who conducts the CNP fit test.

(7) The QNFT protocol shall be followed according to Section I.C.1 of this appendix with an exception for the CNP test exercises.

(b) *CNP test exercises:*

(1) *Normal breathing:* In a normal standing position, without talking, the subject shall breathe normally for 1 minute. After the normal breathing exercise, the subject needs to hold head straight ahead and hold his or her breath for 10 seconds during the test measurement.

(2) *Deep breathing:* In a normal standing position, the subject shall breathe slowly and deeply for 1 minute, being careful not to hyperventilate. After the deep breathing exercise, the subject needs to hold head straight ahead and hold his or her breath for 10 seconds during the test measurement.

(3) *Turning head side to side:* Standing in place, the subject shall slowly turn his or her head from side to side between the extreme positions on each side for 1 minute. The head

shall be held at each extreme momentarily so the subject can inhale at each side. After the turning head side to side exercise, the subject needs to hold head full left and hold his or her breath for 10 seconds during test measurement. Next, the subject needs to hold head full right and hold his or her breath for 10 seconds during the test measurement.

(4) *Moving head up and down:* Standing in place, the subject shall slowly move his or her head up and down for 1 minute. The subject shall be instructed to inhale in the up position (i.e., when looking toward the ceiling). After the moving head up and down exercise, the subject needs to hold head full up and hold his or her breath for 10 seconds during the test measurement. Next, the subject shall hold his or her head full down and hold his or her breath for 10 seconds during the test measurement.

(5) *Talking:* The subject shall talk out loud slowly and loud enough so as to be heard clearly by the test conductor. The subject can read from a prepared text such as the "Rainbow Passage," count backward from 100, or recite a memorized poem or song for 1 minute. After the talking exercise, the subject needs to hold head straight ahead and hold his or her breath for 10 seconds during the test measurement.

(6) *Grimace:* The test subject shall grimace by smiling or frowning for 15 seconds.

(7) *Bending over:* The test subject shall bend at the waist as if he or she were to touch his or her toes for 1 minute. Jogging in place shall be substituted for this exercise in those test environments such as shroud-type QNFT units that prohibit bending at the waist. After the bending over exercise, the subject needs to hold head straight ahead and hold his or her breath for 10 seconds during the test measurement.

(8) *Normal breathing:* The test subject shall remove and re-don the respirator within a 1-minute period. Then, in a normal standing position, without talking, the subject shall breathe normally for 1 minute. After the normal breathing exercise, the subject needs to hold head straight ahead and hold his or her breath for 10 seconds during the test measurement. After the test exercises, the test subject shall be questioned by the test conductor regarding the comfort of the respirator upon completion of the protocol. If it has become unacceptable, another model of a respirator shall be tried.

(c) *CNP test instrument:*

(1) The test instrument shall have an effective audio warning device when the test subject fails to hold his or her breath during the test. The test shall be terminated whenever the test subject has failed to hold his or her breath. The test subject may be refitted and retested.

(2) A record of the test shall be kept on file, assuming the fit test was successful. The record must contain the test subject's name; overall fit factor; make, model, style, and size of respirator used; and date tested.

## *Part II. New fit test protocols*

A. Any person may submit to OSHA an application for approval of a new fit test protocol. If the application meets the following criteria, OSHA will initiate a rulemaking proceeding under Section 6(b)(7) of the OSH Act to determine whether to list the new protocol as an approved protocol in this Appendix A.

B. The application must include a detailed description of the proposed new fit test protocol. This application must be supported by either:

1. A test report prepared by an independent government research laboratory (e.g., Lawrence Livermore National Laboratory, Los Alamos National Laboratory, the National Institute for Standards and Technology) stating that the laboratory has tested the protocol and has found it to be accurate and reliable; or

2. An article that has been published in a peer-reviewed industrial hygiene journal describing the protocol and explaining how test data support the protocol's accuracy and reliability.

C. If OSHA determines that additional information is required before the Agency commences a rulemaking proceeding under this section, OSHA will so notify the applicant and afford the applicant the opportunity to submit the supplemental information. Initiation of a rulemaking proceeding will be deferred until OSHA has received and evaluated the supplemental information. [63 FR 20098, April 23, 1998]

# *Appendix B-1 to §1910.134: user seal check procedures (mandatory)*

The individual who uses a tight-fitting respirator is to perform a user seal check to ensure that an adequate seal is achieved each time the respirator is put on. Either the positive and negative pressure checks listed in this appendix or the respirator manufacturer's recommended user seal check method shall be used. User seal checks are not substitutes for qualitative or quantitative fit tests.

I. *Facepiece positive and/or negative pressure checks:*

A. *Positive pressure check:* Close off the exhalation valve and exhale gently into the facepiece. The face fit is considered satisfactory if a slight positive pressure can be built up inside the facepiece without any evidence of outward leakage of air at the seal. For most respirators, this method of leak testing requires the wearer to first remove the exhalation valve cover before closing off the exhalation valve and then carefully replacing it after the test.

B. *Negative pressure check:* Close off the inlet opening of the canister or cartridge(s) by covering with the palm of the hand(s) or by replacing the filter seal(s), inhale gently so that the facepiece collapses slightly, and hold the breath for 10 seconds. The design of the inlet opening of some cartridges cannot be effectively covered with the palm of the hand. The test can be performed by covering the inlet opening of the cartridge with a thin latex or nitrile glove. If the facepiece remains in its slightly collapsed condition and no inward leakage of air is detected, the tightness of the respirator is considered satisfactory.

II. *Manufacturer's recommended user seal check procedures:* The respirator manufacturer's recommended procedures for performing a user seal check may be used instead of the positive and/or negative pressure check procedures provided that the employer demonstrates that the manufacturer's procedures are equally effective. [63 FR 1152, Jan. 8, 1998]

## *Appendix B-2 to §1910.134: respirator cleaning procedures (mandatory)*

These procedures are provided for employer use when cleaning respirators. They are general in nature, and the employer as an alternative may use the cleaning recommendations provided by the manufacturer of the respirators used by their employees, provided such procedures are as effective as those listed here in Appendix B-2. Equivalent effectiveness simply means that the procedures used must accomplish the objectives set forth in Appendix B-2 (i.e., must ensure that the respirator is properly cleaned and disinfected in a manner that prevents damage to the respirator and does not cause harm to the user).

I. *Procedures for cleaning respirators:*

A. Remove filters, cartridges, or canisters. Disassemble facepieces by removing speaking diaphragms, demand and pressure-demand valve assemblies, hoses, or any components recommended by the manufacturer. Discard or repair any defective parts.

B. Wash components in warm water (43 deg. C [110 deg. F] maximum) with a mild detergent or with a cleaner recommended by the manufacturer. A stiff bristle (not wire) brush may be used to facilitate the removal of dirt.

C. Rinse components thoroughly in clean, warm (43 deg. C [110 deg. F] maximum), preferably running water. Drain.

D. When the cleaner used does not contain a disinfecting agent, respirator components should be immersed for 2 minutes in one of the following:

1. Hypochlorite solution (50 ppm of chlorine) made by adding approximately 1 ml of laundry bleach to 1 liter of water at 43 deg. C (110 deg. F); or,

2. Aqueous solution of iodine (50 ppm iodine) made by adding approximately 0.8 ml of tincture of iodine (6–8 grams ammonium and/or potassium iodide/100 cc of 45% alcohol) to 1 liter of water at 43 deg. C (110 deg. F); or,

3. Other commercially available cleansers of equivalent disinfectant quality when used as directed, if their use is recommended or approved by the respirator manufacturer.

E. Rinse components thoroughly in clean, warm (43 deg. C [110 deg. F] maximum), preferably running water. Drain. The importance of thorough rinsing cannot be overemphasized. Detergents or disinfectants that dry on facepieces may result in dermatitis. In addition, some disinfectants may cause deterioration of rubber or corrosion of metal parts if not completely removed.

F. Components should be hand-dried with a clean lint-free cloth or air-dried.

G. Reassemble facepiece, replacing filters, cartridges, and canisters where necessary.

H. Test the respirator to ensure that all components work properly. [63 FR 1152, Jan. 8, 1998]

## *Appendix C to §1910.134: OSHA respirator medical evaluation questionnaire (mandatory)*

*To the employer:* Answers to questions in Section 1 and to question 9 in Section 2 of Part A do not require a medical examination.

*To the employee:* Can you read (circle one)? *Yes/No*

Your employer must allow you to answer this questionnaire during normal working hours, or at a time and place that is convenient to you. To maintain your confidentiality, your employer or supervisor must not look at or review your answers, and your employer must tell you how to deliver or send this questionnaire to the healthcare professional who will review it.

## *Part A. Section 1 (mandatory)*

The following information must be provided by every employee who has been selected to use any type of respirator (please print):

1. Today's date:_______________________________________________
2. Your name: ________________________________________________
3. Your age (to nearest year):___________________________________
4. Sex (circle one): Male/Female
5. Your height: __________ ft. __________ in.
6. Your weight: __________ lbs.
7. Your job title: _____________________________________________
8. A phone number where you can be reached by the healthcare professional who reviews this questionnaire (include the Area Code):

   ____________________________________________

9. The best time to phone you at this number: ____________________
10. Has your employer told you how to contact the healthcare professional who will review this questionnaire (circle one)? *Yes/No*
11. Check the type of respirator you will use (you can check more than one category):

    a. ______ N, R, or P disposable respirator (filter-mask, non-cartridge type only).

    b. ______ Other type (for example, half- or full-facepiece type, powered-air purifying, supplied-air, self-contained breathing apparatus).

12. Have you worn a respirator (circle one)? *Yes/No*

    If "yes," what type(s):_________________________________________

    ___________________________________________________________

## *Part A. Section 2 (mandatory)*

Questions 1 through 9 below must be answered by every employee who has been selected to use any type of respirator (please circle *Yes* or *No*).

1. Do you currently smoke tobacco, or have you smoked tobacco in the last month? *Yes/No*
2. Have you ever had any of the following conditions?

   a. Seizures (fits): *Yes/No*

   b. Diabetes (sugar disease): *Yes/No*

   c. Allergic reactions that interfere with your breathing: *Yes/No*

   d. Claustrophobia (fear of closed-in places): *Yes/No*

   e. Trouble smelling odors: *Yes/No*

3. Have you ever had any of the following pulmonary or lung problems?
    a. Asbestosis: *Yes/No*
    b. Asthma: *Yes/No*
    c. Chronic bronchitis: *Yes/No*
    d. Emphysema: *Yes/No*
    e. Pneumonia: *Yes/No*
    f. Tuberculosis: *Yes/No*
    g. Silicosis: *Yes/No*
    h. Pneumothorax (collapsed lung): *Yes/No*
    i. Lung cancer: *Yes/No*
    j. Broken ribs: *Yes/No*
    k. Any chest injuries or surgeries: *Yes/No*
    l. Any other lung problem that you've been told about: *Yes/No*
4. Do you currently have any of the following symptoms of pulmonary or lung illness?
    a. Shortness of breath: *Yes/No*
    b. Shortness of breath when walking fast on level ground or walking up a slight hill or incline: *Yes/No*
    c. Shortness of breath when walking with other people at an ordinary pace on level ground: *Yes/No*
    d. Have to stop for breath when walking at your own pace on level ground: *Yes/No*
    e. Shortness of breath when washing or dressing yourself: *Yes/No*
    f. Shortness of breath that interferes with your job: *Yes/No*
    g. Coughing that produces phlegm (thick sputum): *Yes/No*
    h. Coughing that wakes you early in the morning: *Yes/No*
    i. Coughing that occurs mostly when you are lying down: *Yes/No*
    j. Coughing up blood in the last month: *Yes/No*
    k. Wheezing: *Yes/No*
    l. Wheezing that interferes with your job: *Yes/No*
    m. Chest pain when you breathe deeply: *Yes/No*
    n. Any other symptoms that you think may be related to lung problems: *Yes/No*
5. Have you ever had any of the following cardiovascular or heart problems?
    a. Heart attack: *Yes/No*
    b. Stroke: *Yes/No*
    c. Angina: *Yes/No*

   d. Heart failure: *Yes/No*
   e. Swelling in your legs or feet (not caused by walking): *Yes/No*
   f. Heart arrhythmia (heart beating irregularly): *Yes/No*
   g. High blood pressure: *Yes/No*
   h. Any other heart problem that you've been told about: *Yes/No*

6. Have you ever had any of the following cardiovascular or heart symptoms?
   a. Frequent pain or tightness in your chest: *Yes/No*
   b. Pain or tightness in your chest during physical activity: *Yes/No*
   c. Pain or tightness in your chest that interferes with your job: *Yes/No*
   d. In the past two years, have you noticed your heart skipping or missing a beat: *Yes/No*
   e. Heartburn or indigestion that is not related to eating: *Yes/No*
   f. Any other symptoms that you think may be related to heart or circulation problems: *Yes/No*
7. Do you currently take medication for any of the following problems?
   a. Breathing or lung problems: *Yes/No*
   b. Heart trouble: *Yes/No*
   c. Blood pressure: *Yes/No*
   d. Seizures (fits): *Yes/No*
8. If you've used a respirator, have you ever had any of the following problems? (If you've never used a respirator, check the following space [ ] and go to question 9.)
   a. Eye irritation: *Yes/No*
   b. Skin allergies or rashes: *Yes/No*
   c. Anxiety: *Yes/No*
   d. General weakness or fatigue: *Yes/No*
   e. Any other problem that interferes with your use of a respirator: *Yes/No*
9. Would you like to talk to the healthcare professional who will review this questionnaire about your answers to this questionnaire? *Yes/No*

Questions 10 to 15 below must be answered by every employee who has been selected to use either a full-facepiece respirator or a self-contained breathing apparatus (SCBA). For employees who have been selected to use other types of respirators, answering these questions is voluntary.

10. Have you ever lost vision in either eye (temporarily or permanently)? *Yes/No*

11. Do you currently have any of the following vision problems?
    a. Wear contact lenses: *Yes/No*
    b. Wear glasses: *Yes/No*
    c. Color blind: *Yes/No*
    d. Any other eye or vision problem: *Yes/No*
12. Have you ever had an injury to your ears, including a broken ear drum? *Yes/No*
13. Do you currently have any of the following hearing problems?
    a. Difficulty hearing: *Yes/No*
    b. Wear a hearing aid: *Yes/No*
    c. Any other hearing or ear problem: *Yes/No*
14. Have you ever had a back injury? *Yes/No*
15. Do you currently have any of the following musculoskeletal problems?
    a. Weakness in any of your arms, hands, legs, or feet: *Yes/No*
    b. Back pain: *Yes/No*
    c. Difficulty fully moving your arms and legs: *Yes/No*
    d. Pain or stiffness when you lean forward or backward at the waist: *Yes/No*
    e. Difficulty fully moving your head up or down: *Yes/No*
    f. Difficulty fully moving your head side to side: *Yes/No*
    g. Difficulty bending at your knees: *Yes/No*
    h. Difficulty squatting to the ground: *Yes/No*
    i. Climbing a flight of stairs or a ladder carrying more than 25 lbs.: *Yes/No*
    j. Any other muscle or skeletal problem that interferes with using a respirator: *Yes/No*

## *Part B*

Any of the following questions, and other questions not listed, may be added to the questionnaire at the discretion of the healthcare professional who will review the questionnaire.

1. In your current job, are you working at high altitudes (over 5000 feet) or in a place that has lower than normal amounts of oxygen? *Yes/No*

   If "yes," do you have feelings of dizziness, shortness of breath, pounding in your chest, or other symptoms when you're working under these conditions? *Yes/No*
2. At work or at home, have you ever been exposed to hazardous solvents or hazardous airborne chemicals (e.g., gases, fumes, or dust) or have you come into skin contact with hazardous chemicals? *Yes/No*

If "yes," name the chemicals if you know them:

______________________________________________

______________________________________________

______________________________________________

______________________________________________

3. Have you ever worked with any of the materials, or under any of the conditions, listed below?
   a. Asbestos: *Yes/No*
   b. Silica (e.g., in sandblasting): *Yes/No*
   c. Tungsten/cobalt (e.g., grinding or welding this material): *Yes/No*
   d. Beryllium: *Yes/No*
   e. Aluminum: *Yes/No*
   f. Coal (for example, mining): *Yes/No*
   g. Iron: *Yes/No*
   h. Tin: *Yes/No*
   i. Dusty environments: *Yes/No*
   j. Any other hazardous exposures: *Yes/No*

   If "yes," describe these exposures:

______________________________________________

______________________________________________

______________________________________________

______________________________________________

4. List any second jobs or side businesses you have:

______________________________________________

______________________________________________

______________________________________________

______________________________________________

5. List your previous occupations:

______________________________________________

______________________________________________

______________________________________________

______________________________________________

6. List your current and previous hobbies:

______________________________________________

______________________________________________

______________________________________________

______________________________________________

7. Have you been in the military services? *Yes/No*

   If "yes," were you exposed to biological or chemical agents (either in training or combat)? *Yes/No*

8. Have you ever worked on a HAZMAT team? *Yes/No*

9. Other than medications for breathing and lung problems, heart trouble, blood pressure, and seizures mentioned earlier in this questionnaire, are you taking any other medications for any reason (including over-the-counter medications)? *Yes/No*

   If "yes," name the medications if you know them:

   ____________________________________________

   ____________________________________________

   ____________________________________________

   ____________________________________________

10. Will you be using any of the following items with your respirator(s)?

    a. HEPA filters: *Yes/No*

    b. Canisters (for example, gas masks): *Yes/No*

    c. Cartridges: *Yes/No*

11. How often are you expected to use the respirator(s) (circle *Yes* or *No* for all answers that apply to you)?

    a. Escape only (no rescue): *Yes/No*

    b. Emergency rescue only: *Yes/No*

    c. Less than 5 hours per week: *Yes/No*

    d. Less than 2 hours per day: *Yes/No*

    e. 2 to 4 hours per day: *Yes/No*

    f. Over 4 hours per day: *Yes/No*

12. During the period you are using the respirator(s), is your work effort:

    a. Light (less than 200 kcal per hour): *Yes/No*

       If "yes," how long does this period last during the average shift? ___________hrs. ___________mins.

       Examples of a light work effort are sitting while writing, typing, drafting, or performing light assembly work; or standing while operating a drill press (1–3 lbs.) or controlling machines.

    b. Moderate (200 to 350 kcal per hour): *Yes/No*

       If "yes," how long does this period last during the average shift? ___________hrs. ___________mins.

       Examples of moderate work effort are sitting while nailing or filing; driving a truck or bus in urban traffic; standing while drilling, nailing, performing assembly work, or transferring a moderate load (about 35 lbs.) at trunk level; walking on a level surface about 2 mph or down a 5-degree grade about 3 mph; or pushing a wheelbarrow with a heavy load (about 100 lbs.) on a level surface.

c. Heavy (above 350 kcal per hour): *Yes/No*

If "yes," how long does this period last during the average shift? ___________hrs. ___________mins.

Examples of heavy work are lifting a heavy load (about 50 lbs.) from the floor to your waist or shoulder; working on a loading dock; shoveling; standing while bricklaying or chipping castings; walking up an 8-degree grade about 2 mph; climbing stairs with a heavy load (about 50 lbs.).

13. Will you be wearing protective clothing and/or equipment (other than the respirator) when you're using your respirator? *Yes/No*

    If "yes," describe this protective clothing and/or equipment:

    ___________________________________________

    ___________________________________________

    ___________________________________________

    ___________________________________________

14. Will you be working under hot conditions (temperature exceeding 77 deg. F)? *Yes/No*
15. Will you be working under humid conditions? *Yes/No*
16. Describe the work you'll be doing while you're using your respirator(s):

    ___________________________________________

    ___________________________________________

    ___________________________________________

    ___________________________________________

17. Describe any special or hazardous conditions you might encounter when you're using your respirator(s) (for example, confined spaces, life-threatening gases):

    ___________________________________________

    ___________________________________________

    ___________________________________________

    ___________________________________________

18. Provide the following information, if you know it, for each toxic substance that you'll be exposed to when you're using your respirator(s):

    Name of the first toxic substance: ___________________________

    Estimated maximum exposure level per shift:___________________

    Duration of exposure per shift:______________________________

    Name of the second toxic substance:__________________________

Estimated maximum exposure level per shift:____________________

Duration of exposure per shift:____________________

Name of the third toxic substance:____________________

Estimated maximum exposure level per shift:____________________

Duration of exposure per shift:____________________

The name of any other toxic substances that you'll be exposed to while using your respirator:

____________________

____________________

____________________

____________________

19. Describe any special responsibilities you'll have while using your respirator(s) that may affect the safety and well-being of others (for example, rescue or security):

____________________

____________________

____________________

____________________

*appendix D*

# *Confined spaces*

## *29 C.F.R. 1910.146 Appendix A: PRCS decision flow chart*

See next page.

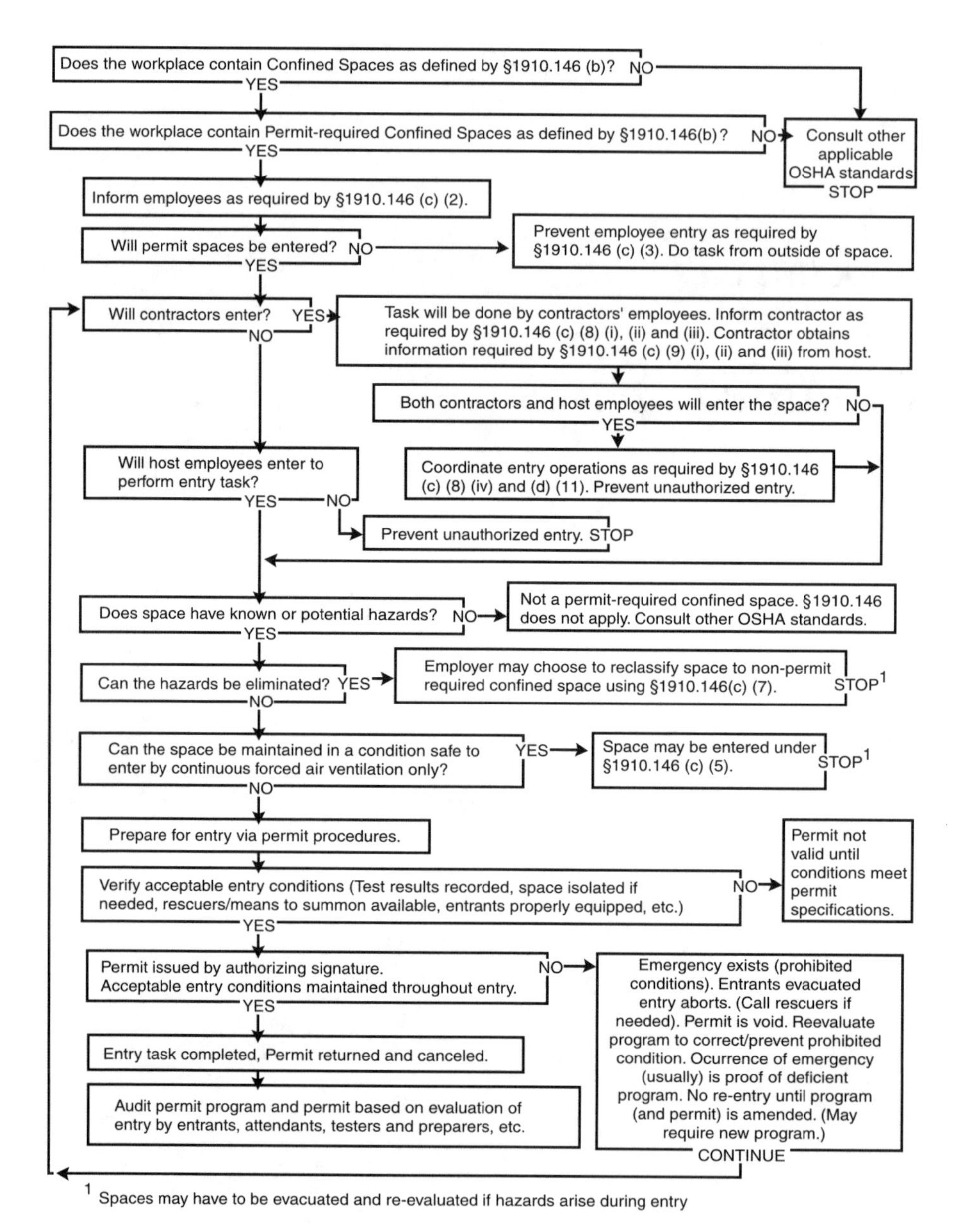


29 C.F.R. 1910.146 Appendix A: Permit-required confined-space decision flow chart.

# *29 C.F.R. 1910.146 Appendix B: procedures for atmospheric testing*

*Subpart number: J*
*Subpart title: General Environmental Controls*

Atmospheric testing is required for two distinct purposes: (1) evaluation of the hazards of the permit space, and (2) verification that acceptable entry conditions for entry into that space exist.

## *Evaluation testing*

The atmosphere of a confined space should be analyzed using equipment of sufficient sensitivity and specificity to identify and evaluate any hazardous atmospheres that may exist or arise, so that appropriate permit entry procedures can be developed and acceptable entry conditions stipulated for that space. Evaluation and interpretation of these data, and development of the entry procedure, should be done by, or reviewed by, a technically qualified professional (e.g., OSHA consultation service or certified industrial hygienist, registered safety engineer, certified safety professional, certified marine chemist, etc.) based on evaluation of all serious hazards.

## *Verification testing*

The atmosphere of a permit space which may contain a hazardous atmosphere should be tested for residues of all contaminants identified by evaluation testing using permit-specified equipment to determine that residual concentrations at the time of testing and entry are within the range of acceptable entry conditions. Results of testing (e.g., actual concentration, etc.) should be recorded on the permit in the space provided adjacent to the stipulated acceptable entry condition.

## *Duration of testing*

Measurement of values for each atmospheric parameter should be made for at least the minimum response time of the test instrument specified by the manufacturer.

## *Testing stratified atmospheres*

When monitoring for entries involving a descent into atmospheres that may be stratified, the atmospheric envelope should be tested a distance of approximately 4 feet (1.22 m) in the direction of travel and to each side. If a sampling probe is used, the entrant's rate of progress should be slowed to accommodate the sampling speed and detector response.

### Order of testing

A test for oxygen is performed first because most combustible gas meters are oxygen dependent and will not provide reliable readings in an oxygen-deficient atmosphere. Combustible gases are tested for next because the threat of fire or explosion is both more immediate and more life threatening, in most cases, than exposure to toxic gases and vapors. If tests for toxic gases and vapors are necessary, they are performed last.

## 29 C.F.R. 1910.146 Appendix D: sample permits

### Appendix D-1: confined space entry permit

Date and time issued: ________________

Date and time expires: ________________

Job site/space I.D.: ________________

Job supervisor: ________________

Equipment to be worked on: ________________

Work to be performed: ________________

Standby personnel: ________ ________ ________

1. Atmospheric checks:

    Time ________
    Oxygen ________%
    Explosive ________% L.F.L.
    Toxic ________ ppm

2. Tester's signature: ________________
3. Source isolation (no entry):

    Pumps or lines blinded, disconnected, or blocked?

    N/A [ ] Yes [ ] No [ ]

4. Ventilation modification:

    Mechanical?

    N/A [ ] Yes [ ] No [ ]

    Natural ventilation only?

    N/A [ ] Yes [ ] No [ ]

5. Atmospheric check after isolation and ventilation:

    Time ________
    Oxygen ________% > 19.5 %
    Explosive ________% L.F.L. < 10 %
    Toxic ________ ppm < 10 ppm $H_2S$

    Tester's signature: ________________

6. Communication procedures:

___________________________________________

___________________________________________

7. Rescue procedures:

___________________________________________

___________________________________________

| 8. Entry, standby, and backup persons: | | Yes | No |
|---|---|---|---|
| Successfully completed required training? | | [ ] | [ ] |
| Is it current? | | [ ] | [ ] |
| **9. Equipment:** | N/A | Yes | No |
| Direct reading gas monitor tested | [ ] | [ ] | [ ] |
| Safety harnesses and lifelines for entry and standby persons | [ ] | [ ] | [ ] |
| Hoisting equipment | [ ] | [ ] | [ ] |
| Powered communications | [ ] | [ ] | [ ] |
| SCBAs for entry and standby persons | [ ] | [ ] | [ ] |
| Protective clothing | [ ] | [ ] | [ ] |
| All electric equipment listed Class I, Division I, Group D, and non-sparking tools | [ ] | [ ] | [ ] |

10. Periodic atmospheric tests:

Oxygen ______% Time ______ Oxygen ______% Time ______
Oxygen ______% Time ______ Oxygen ______% Time ______

Explosive ______% Time ______ Explosive ______% Time ______
Explosive ______% Time ______ Explosive ______% Time ______

Toxic ______% Time ______ Toxic ______% Time ______
Toxic ______% Time ______ Toxic ______% Time ______

We have reviewed the work authorized by this permit and the information contained herein. Written instructions and safety procedures have been received and are understood. Entry cannot be approved if any squares are marked in the "No" column. This permit is not valid unless all appropriate items are completed.

Permit prepared by (supervisor):___________________________

Approved by (unit supervisor): ___________________________

Reviewed by (operations personnel):

___________________________ ___________________________
(printed name) (signature)

*This permit to be kept at jobsite. Return jobsite copy to safety office following job completion.*

Copies: White original (safety office) Yellow (unit supervisor) Hard (jobsite)

## *Appendix D-2: entry permit*

Permit valid for 8 hours only. All copies of permit will remain at jobsite until job is completed

Date: _____-_____-_____

Site location and description ____________________________________

Purpose of entry ____________________________________________

| Supervisor(s) in charge of crews | Type of Crew | Phone # |
|---|---|---|
| ____________________ | ______________ | __________ |
| ____________________ | ______________ | __________ |
| ____________________ | ______________ | __________ |

Communication procedures:

____________________________________________________________

____________________________________________________________

____________________________________________________________

Rescue procedures (phone numbers at bottom):

____________________________________________________________

____________________________________________________________

____________________________________________________________

Bold denotes minimum requirements to be completed and reviewed prior to entry.

| Requirements Completed | Date | Time |
|---|---|---|
| Lockout/de-energize/tryout | ________ | ________ |
| Line(s) broken-capped-blanked | ________ | ________ |
| Purge-flush and vent | ________ | ________ |
| Ventilation | ________ | ________ |
| Secure area (post and flag) | ________ | ________ |
| Breathing apparatus | ________ | ________ |
| Resuscitator/inhalator | ________ | ________ |
| Standby safety personnel | ________ | ________ |
| Full body harness with D-ring | ________ | ________ |
| Emergency escape retrieval equipment | ________ | ________ |
| Lifelines | ________ | ________ |
| Fire extinguishers | ________ | ________ |
| Lighting (explosive proof) | ________ | ________ |
| Protective clothing | ________ | ________ |
| Respirator(s) (air purifying) | ________ | ________ |
| Burning and welding permit | ________ | ________ |

*Note:* For items that do not apply, enter N/A.

Record continuous monitoring results every 2 hours.

*Continuous monitoring:*

| Test(s) To Be Taken | Permissible Entry Level | | | | | | |
|---|---|---|---|---|---|---|---|
| Percent of oxygen | 19.5% to 23.5% | ____ | ____ | ____ | ____ | ____ | ____ |
| Lower flammable limit | Under 10% | ____ | ____ | ____ | ____ | ____ | ____ |
| Carbon monoxide | 35[1] ppm | ____ | ____ | ____ | ____ | ____ | ____ |
| Aromatic hydrocarbon | 1 ppm[1]/5 ppm[2] | ____ | ____ | ____ | ____ | ____ | ____ |
| Hydrogen cyanide (skin) | 4 ppm[2] | ____ | ____ | ____ | ____ | ____ | ____ |
| Hydrogen sulfide | 10 ppm[1]/15 ppm[2] | ____ | ____ | ____ | ____ | ____ | ____ |
| Sulfur dioxide | 2 ppm[1]/5 ppm[2] | ____ | ____ | ____ | ____ | ____ | ____ |
| Ammonia | 35 ppm[2] | ____ | ____ | ____ | ____ | ____ | ____ |

[1] 8-hour, time-weighted average — employee can work in area 8 hours (longer with appropriate respiratory protection).

[2] Short-term exposure limit — employee can work in the area up to 15 minutes.

Remarks: ____________________________________________

| Gas Tester Name and Check # | Instrument(s) Used | Model and/or Type | Serial and/or Unit # |
|---|---|---|---|
| ____________ | ____________ | ____________ | ____________ |
| ____________ | ____________ | ____________ | ____________ |
| ____________ | ____________ | ____________ | ____________ |

Safety standby person is required for all confined space work.

| Safety Standby Person | Check # | Confined Space Entrant(s) | Check # | Confined Space Entrant(s) | Check # |
|---|---|---|---|---|---|
| __________ | ____ | __________ | ____ | __________ | ____ |
| __________ | ____ | __________ | ____ | __________ | ____ |
| __________ | ____ | __________ | ____ | __________ | ____ |

Supervisor authorizing that all conditions are satisfied:

____________________________________________

Department/phone: ______________________

Ambulance/2800; Fire/2900; Safety/4901; Gas Coordinator/4529/5387.

[58 FR 4549, Jan. 14, 1993; 58 FR 34846, June 29, 1993]

# *appendix E*

# *Fall hazards**

## *Inspection and maintenance*

To maintain their service life and high performance, all belts and harnesses should be inspected frequently. Visual inspection before each use should become routine, as well as routine inspection by a competent person. If any of the conditions listed below are found, the equipment should be replaced before being used.

## *Harness inspection*

### *Belts and rings*

For harness inspections, begin at one end and hold the body side of the belt toward you, grasping the belt with your hands 6 to 8 inches apart. Bend the belt in an inverted "U." Watch for frayed edges, broken fibers, pulled stitches, cuts, or chemical damage. Check D-rings and D-ring metal wear pads for distortion, cracks, breaks, and rough or sharp edges. The D-ring bar should be at a 90° angle to the long axis of the belt and should pivot freely. Attachments of buckles and D-rings should be given special attention. Note any unusual wear, frayed or cut fibers, or distortion of the buckles. Rivets should be tight and not be removable with fingers. The body-side rivet base and outside rivets should be flat against the material. Bent rivets will fail under stress. Inspect frayed or broken strands. Broken webbing strands generally appear as tufts on the webbing surface. Any broken, cut, or burned stitches will be readily seen.

* From http://www.osha-slc.gov/Region/fallprotection/fall_protection_info.html. Information contained within this document was obtained partially from *The St. Paul Tie or Die Fall Protection Program* manual.

### Tongue buckle

Buckle tongues should be free of distortion in shape and motion. They should overlap the buckle frame and move freely back and forth in their socket. Rollers should turn freely on the frame. Check for distortion or sharp edges.

### Friction buckle

Inspect the buckle for distortion. The outer bar or center bars must be straight. Pay special attention to corners and attachment points of the center bar.

## Lanyard inspection

When inspecting lanyards, begin at one end and work to the opposite end. Slowly rotate the lanyard so that the entire circumference is checked. Spliced ends require particular attention. Hardware should be examined using the procedures detailed below.

### Hardware

#### Snaps

Inspect closely for hook and eye distortion, cracks, corrosion, or pitted surfaces. The keeper or latch should seat into the nose without binding and should not be distorted or obstructed. The keeper spring should exert sufficient force to close the keeper firmly. Keeper rocks must prevent the keeper from opening when the keeper is closed.

#### Thimbles

The thimble (protective plastic sleeve) must be firmly seated in the eye of the splice, and the splice should have no loose or cut strands. The edges of the thimble should be free of sharp edges, distortion, or cracks.

### Lanyards

#### Steel lanyards

While rotating a steel lanyard, watch for cuts, frayed areas, or unusual wear patterns on the wire. The use of steel lanyards for fall protection without a shock-absorbing device is not recommended.

#### Web lanyards

While bending webbing over a piece of pipe, observe each side of the webbed lanyard. This will reveal any cuts or breaks. Due to the limited elasticity of the web lanyard, fall protection without the use of a shock absorber is not recommended.

#### *Rope lanyard*

Rotation of the rope lanyard while inspecting it from end to end will bring to light any fuzzy, worn, broken, or cut fibers. Weakened areas from extreme loads will appear as a noticeable change in original diameter. The rope diameter should be uniform throughout, following a short break-in period. When a rope lanyard is used for fall protection, a shock-absorbing system should be included.

### *Visual indication of damage to webbing and rope lanyards*

#### *Heat*

In excessive heat, nylon becomes brittle and has a shriveled, brownish appearance. Fibers will break when flexed and should not be used above 180°F.

#### *Chemical*

Change in color usually appears as a brownish smear or smudge. Transverse cracks, which appear when belt is bent over tight, causes a loss of elasticity in the belt.

#### *Ultraviolet rays*

Do not store webbing and rope lanyards in direct sunlight, because ultraviolet rays can reduce the strength of some material.

#### *Molten metal or flame*

Webbing and rope strands may be fused together by molten metal or flame. Watch for hard, shiny spots or a hard and brittle feel. Webbing will not support combustion, but nylon will.

#### *Paint and solvents*

Paint will penetrate and dry, restricting movements of fibers. Drying agents and solvents in some paints will appear as chemical damage.

## *Shock-absorbing packs*

The outer portion of the shock-absorbing pack should be examined for burn holes and tears. Stitching on areas where the pack is sewn to the D-ring, belt, or lanyard should be examined for loose strands, rips, and deterioration.

## *Cleaning of equipment*

Basic care for fall-protection safety equipment will prolong and endure the life of the equipment and contribute toward the performance of its vital safety function. Proper storage and maintenance after use are as important as cleaning the equipment of dirt, corrosives, or contaminants. The storage area should be clean, dry, and free of exposure to fumes or corrosive elements.

### *Nylon and polyester*

Wipe off all surface dirt with a sponge dampened in plain water. Squeeze the sponge dry. Dip the sponge in a mild solution of water and commercial soap or detergent. Work up a thick lather with a vigorous back-and-forth motion, then wipe the belt dry with a clean cloth. Hang freely to dry but away from excessive heat.

### *Drying*

Harnesses, belts, and other equipment should be dried thoroughly without exposure to heat, steam, or long periods of sunlight.

# *appendix F*

# *29 C.F.R. 1910.1000: air contaminants*

*Subpart number: Z*
*Subpart title: Toxic and Hazardous Substances*

An employee's exposure to any substance listed in Tables Z-1, Z-2, or Z-3 of this section shall be limited in accordance with the requirements of the following paragraphs of this section.

## *29 C.F.R. 1910.1000(a): Table Z-1*

(a)(1) *Substances with limits preceded by "C" — ceiling values:* An employee's exposure to any substance in Table Z-1, the exposure limit of which is followed by a "(C)," shall at no time exceed the exposure limit given for that substance. If instantaneous monitoring is not feasible, then the ceiling shall be assessed as a 15-minute, time-weighted average exposure which shall not be exceeded at any time during the working day.

(a)(2) *Other substances — 8-hour time-weighted averages:* An employee's exposure to any substance in Table Z-1, the exposure limit of which is not followed by a "(C)," shall not exceed the 8-hour, time-weighted average given for that substance any 8-hour work shift of a 40-hour work week.

## *29 C.F.R. 1910.1000(b): Table Z-2*

An employee's exposure to any substance listed in Table Z-2 shall not exceed the exposure limits specified as follows:

(b)(1) *8-hour time-weighted averages:* An employee's exposure to any substance listed in Table Z-2, in any 8-hour work shift of a 40-hour work week, shall not exceed the 8-hour, time-weighted average limit given for that substance in Table Z-2.

(b)(2) *Acceptable ceiling concentrations:* An employee's exposure to a substance listed in Table Z-2 shall not exceed at any time during an 8-hour shift the acceptable ceiling concentration limit given for the substance in the table, except for a time period, and up to a concentration not exceeding the maximum duration and concentration allowed, in the column under "acceptable maximum peak above the acceptable ceiling concentration for an 8-hour shift."

(b)(3) *Example:* During an 8-hour work shift, an employee may be exposed to a concentration of Substance A (with a 10-ppm TWA, 25-ppm ceiling, and 50-ppm peak) above 25 ppm (but never above 50 ppm) only for a maximum period of 10 minutes. Such exposure must be compensated by exposures to concentrations less than 10 ppm so that the cumulative exposure for the entire 8-hour workshift does not exceed a weighted average of 10 ppm.

## 29 C.F.R. 1910.1000(c): Table Z-3

An employee's exposure to any substance listed in Table Z-3, in any 8-hour work shift of a 40-hour work week, shall not exceed the 8-hour, time-weighted average limit given for that substance in the table.

## 29 C.F.R. 1910.1000(d): Computation formulae

The computation formula which shall apply to employee exposure to more than one substance for which 8-hour, time-weighted averages are listed in subpart Z of 29 C.F.R. Part 1910 in order to determine whether an employee is exposed over the regulatory limit is as follows:

(d)(1)

(d)(1)(i) The cumulative exposure for an 8-hour work shift shall be computed as follows:

$$E = (C(a)T(a) + C(b)T(b) + \dots C(n)T(n)) \div 8$$

where:

$E$ is the equivalent exposure for the working shift.

$C$ is the concentration during any period of time $T$ where the concentration remains constant.

$T$ is the duration in hours of the exposure at the concentration $C$.

The value of $E$ shall not exceed the 8-hour, time-weighted average specified in Subpart Z or 29 C.F.R. Part 1910 for the substance involved.

(d)(1)(ii) To illustrate the formula prescribed in paragraph (d)(1)(i) of this section, assume that Substance A has an 8-hour, time-weighted average limit of 100 ppm as noted in Table Z-1. Assume that an employee is subject to the following exposure:

Two hours of exposure at 150 ppm
Two hours of exposure at 75 ppm
Four hours of exposure at 50 ppm

Substituting this information in the formula, we have

$$((2 \times 150) + (2 \times 75) + (4 \times 50)) \div 8 = 81.25 \text{ ppm}$$

Since 81.25 ppm is less than 100 ppm, the 8-hour, time-weighted average limit, the exposure is acceptable.

(d)(2)

(d)(2)(i) In case of a mixture of air contaminants, an employer shall compute the equivalent exposure as follows:

$$E(m) = (C(1) \div L(1)) + (C(2) \div L(2)) + \dots (C(n) \div L(n))$$

where:

$E(m)$ is the equivalent exposure for the mixture.

$C$ is the concentration of a particular contaminant.

$L$ is the exposure limit for that substance specified in Subpart Z of 29 C.F.R. Part 1910.

The value of $E(m)$ shall not exceed unity (1).

(d)(2)(ii) To illustrate the formula prescribed in paragraph (d)(2)(i) of this section, consider the following exposures:

| Substance | Actual Concentration of 8-hr Exposure (ppm) | 8-hr TWA PEL (ppm) |
|---|---|---|
| B | 500 | 1000 |
| C | 45 | 200 |
| D | 40 | 200 |

Substituting in the formula, we have:

$$E(m) = (500 \div 1000) + (45 \div 200) + (40 \div 200)$$
$$E(m) = 0.500 + 0.225 + 0.200$$
$$E(m) = 0.925$$

Since $E(m)$ is less than unity (1), the exposure combination is within acceptable limits.

## 29 C.F.R. 1910.1000(e)

To achieve compliance with paragraphs (a) through (d) of this section, administrative or engineering controls must first be determined and implemented whenever feasible. When such controls are not feasible to achieve full compliance, protective equipment or any other protective measures shall be used to keep the exposure of employees to air contaminants within the limits prescribed in this section. Any equipment and/or technical measures used for this purpose must be approved for each particular use by a competent industrial hygienist or other technically qualified person. Whenever respirators are used, their use shall comply with §1910.134.

## 29 C.F.R. 1910.1000(f): Effective dates

The exposure limits specified have been in effect with the method of compliance specified in paragraph (e) of this section since May 29, 1971.

## Table Z-1. Limits for air contaminants

*Note:* Because of the length of the table, explanatory footnotes applicable to all substances are given below as well as at the end of the table. Footnotes specific only to a limited number of substances are also shown within the table.

Footnote 1: The PELs are 8-hr TWAs unless otherwise noted; a (C) designation denotes a ceiling limit. They are to be determined from breathing-zone air samples.

Footnote a: Parts of vapor or gas per million parts of contaminated air by volume at 25°C and 760 torr.

Footnote b: Milligrams of substance per cubic meter of air. When entry is in this column only, the value is exact; when listed with a ppm entry, it is approximate.

Footnote c: The CAS number is for information only. Enforcement is based on the substance name. For an entry covering more than one metal compound measured as the metal, the CAS number for the metal is given — not CAS numbers for the individual compounds.

Footnote d: The final benzene standard in §1910.1028 applies to all occupational exposures to benzene except in some circumstances the distribution and sale of fuels, sealed containers and pipelines, coke production, oil and gas drilling and production, natural gas processing, and the percentage exclusion for liquid mixtures; for the excepted subsegments, the benzene limits in Table Z-2 apply. See §1910.1028 for specific circumstances.

Footnote e: This 8-hr TWA applies to respirable dust as measured by a vertical elutriator cotton dust sampler or equivalent instrument. The time-weighted average applies to the cotton waste processing operations of waste recycling (sorting, blending, cleaning, and willowing) and garnetting. See also §1910.1043 for cotton dust limits applicable to other sectors.

Footnote f: All inert or nuisance dusts, whether mineral, inorganic, or organic, not listed specifically by substance name are covered by the "particulates not otherwise regulated" (PNOR) limit, which is the same as the inert or nuisance dust limit of Table Z-3.

Footnote 2: See Table Z-2.

Footnote 3: See Table Z-3 (refer to 62 FR 42018, Aug. 4, 1997).

Footnote 4: Varies with compound.

***Table Z-1*** Limits for Air Contaminants

| Substance | CAS No.[c] | PEL (ppm)[1,a] | PEL (mg/m³)[1,b] | Skin Designation |
|---|---|---|---|---|
| Acetaldehyde | 75-07-0 | 200 | 360 | |
| Acetic acid | 64-19-7 | 10 | 5 | |
| Acetic anhydride | 108-24-7 | 5 | 20 | |
| Acetone | 67-64-1 | 1000 | 2400 | |
| Acetonitrile | 75-05-8 | 40 | 70 | |
| 2-Acetylaminofluorene; see §1910.1014 | 53-96-3 | — | — | |
| Acetylene dichloride; see 1,2-dichloroethylene | | | | |
| Acetylene tetrabromide | 79-27-6 | 1 | 14 | |
| Acrolein | 107-02-8 | 0.1 | 0.25 | |
| Acrylamide | 79-06-1 | — | 0.3 | X |
| Acrylonitrile; see §1910.1045 | 107-13-1 | — | — | |
| Aldrin | 309-00-2 | — | 0.25 | X |
| Allyl alcohol | 107-18-6 | 2 | 5 | X |
| Allyl chloride | 107-05-1 | 1 | 3 | |
| Allyl glycidyl ether (AGE) | 106-92-3 | 10[(C)] | 45[(C)] | |
| Allyl propyl disulfide | 2179-59-1 | 2 | 12 | |
| *alpha*-Alumina | 1344-28-1 | — | — | |
| Total dust | — | — | 15 | |
| Respirable fraction | — | — | 5 | |
| Aluminum metal (as Al) | 7429-90-5 | — | — | |
| Total dust | — | — | 15 | |
| Respirable fraction | — | — | 5 | |
| 4-Aminodiphenyl; see §1910.1011 | 92-67-1 | — | — | |
| 2-Aminoethanol; see ethanolamine | | | | |
| 2-Aminopyridine | 504-29-0 | 0.5 | 2 | |
| Ammonia | 7664-41-7 | 50 | 35 | |
| Ammonium sulfamate | 7773-06-0 | — | — | |
| Total dust | — | — | 15 | |
| Respirable fraction | — | — | 5 | |
| *n*-Amyl acetate | 628-63-7 | 100 | 525 | |
| *sec*-Amyl acetate | 626-38-0 | 125 | 650 | |
| Aniline and homologs | 62-53-3 | 5 | 19 | X |
| Anisidine (*o*-, *p*- isomers) | 29191-52-4 | — | 0.5 | X |
| Antimony and compounds (as Sb) | 7440-36-0 | — | 0.5 | |

**Table Z-1** Limits for Air Contaminants *(continued)*

| Substance | CAS No.[c] | PEL (ppm)[1,a] | PEL (mg/m³)[1,b] | Skin Designation |
|---|---|---|---|---|
| ANTU (*alpha*-naphthylthiourea) | 86-88-4 | — | 0.3 | |
| Arsenic, inorganic compounds (as As); see §1910.1018 | 7440-38-2 | — | — | |
| Arsenic, organic compounds (as As) | 7440-38-2 | — | 0.5 | |
| Arsine | 7784-42-1 | 0.05 | 0.2 | |
| Asbestos; see §1910.1001 | —[4] | — | — | |
| Azinphos-methyl | 86-50-0 | — | 0.2 | X |
| Barium, soluble compounds (as Ba) | 7440-39-3 | — | 0.5 | |
| Barium sulfate | 7727-43-7 | | | |
| Total dust | — | — | 15 | |
| Respirable fraction | — | — | 5 | |
| Benomyl | 17804-35-2 | — | — | |
| Total dust | — | — | 15 | |
| Respirable fraction | — | — | 5 | |
| Benzene; see §1910.1028. See Table Z-2 for the limits applicable in the operations or sectors excluded in §1910.1028[d] | 71-43-2 | — | — | |
| Benzidine; see §1910.1010 | 92-87-5 | — | — | |
| *p*-Benzoquinone; see quinone | | | | |
| Benzo(*a*)pyrene; see coal tar pitch volatiles | | | | |
| Benzoyl peroxide | 94-36-0 | — | 5 | |
| Benzyl chloride | 100-44-7 | 1 | 5 | |
| Beryllium and beryllium compounds (as Be) | 7440-41-7 | — | —[2] | |
| Biphenyl; see diphenyl | | | | |
| Bismuth telluride, undoped | 1304-82-1 | — | — | |
| Total dust | — | — | 15 | |
| Respirable fraction | — | — | 5 | |
| Boron oxide | 1303-86-2 | — | — | |
| Total dust | — | — | 15 | |
| Boron trifluoride | 7637-07-2 | 1[(C)] | 3[(C)] | |
| Bromine | 7726-95-6 | 0.1 | 0.7 | |

***Table Z-1*** Limits for Air Contaminants *(continued)*

| Substance | CAS No.[c] | PEL (ppm)[1,a] | PEL (mg/m³)[1,b] | Skin Designation |
|---|---|---|---|---|
| Bromoform | 75-25-2 | 0.5 | 5 | X |
| Butadiene (1,3-butadiene); see §1910.1051; 1910.19[1] | 106-99-0 | 1 ppm/ 5 ppm STEL | — | |
| Butanethiol; see butyl mercaptan | | | | |
| 2-Butanone (methyl ethyl ketone) | 78-93-3 | 200 | 590 | |
| 2-Butoxyethanol | 111-76-2 | 50 | 240 | X |
| *n*-Butyl-acetate | 123-86-4 | 150 | 710 | |
| *sec*-Butyl-acetate | 105-46-4 | 200 | 950 | |
| *tert*-Butyl-acetate | 540-88-5 | 200 | 950 | |
| *n*-Butyl alcohol | 71-36-3 | 100 | 300 | |
| *sec*-Butyl alcohol | 78-92-2 | 150 | 450 | |
| *tert*-Butyl alcohol | 75-65-0 | 100 | 300 | |
| Butylamine | 109-73-9 | 5[(C)] | 15[(C)] | X |
| *tert*-Butyl chromate (as $CrO_3$) | 1189-85-1 | — | 0.1[(C)] | X |
| *n*-Butyl glycidyl ether (BGE) | 2426-08-6 | 50 | 270 | |
| Butyl mercaptan | 109-79-5 | 10 | 35 | |
| *p-tert*-Butyltoluene | 98-51-1 | 10 | 60 | |
| Cadmium (as Cd); see §1910.1027 | 7440-43-9 | — | — | |
| Calcium carbonate | 1317-65-3 | — | — | |
| Total dust | — | — | 15 | |
| Respirable fraction | — | — | 5 | |
| Calcium hydroxide | 1305-62-0 | — | — | |
| Total dust | — | — | 15 | |
| Respirable fraction | — | — | 5 | |
| Calcium oxide | 1305-78-8 | — | 5 | |
| Calcium silicate | 1344-95-2 | — | — | |
| Total dust | — | — | 15 | |
| Respirable fraction | — | — | 5 | |
| Calcium sulfate | 7778-18-9 | — | — | |
| Total dust | — | — | 15 | |
| Respirable fraction | — | — | 5 | |
| Camphor, synthetic | 76-22-2 | — | 2 | |
| Carbaryl (Sevin) | 63-25-2 | — | 5 | |
| Carbon black | 1333-86-4 | — | 3.5 | |
| Carbon dioxide | 124-38-9 | 5000 | 9000 | |

***Table Z-1*** Limits for Air Contaminants *(continued)*

| Substance | CAS No.[c] | PEL (ppm)[1,a] | PEL (mg/m$^3$)[1,b] | Skin Designation |
|---|---|---|---|---|
| Carbon disulfide | 75-15-0 | — | —[2] | |
| Carbon monoxide | 630-08-0 | 50 | 55 | |
| Carbon tetrachloride | 56-23-5 | — | —[2] | |
| Cellulose | 9004-34-6 | — | — | |
| Total dust | — | — | 15 | |
| Respirable fraction | — | — | 5 | |
| Chlordane | 57-74-9 | — | 0.5 | X |
| Chlorinated camphene | 8001-35-2 | — | 0.5 | X |
| Chlorinated diphenyl oxide | 55720-99-5 | — | 0.5 | |
| Chlorine | 7782-50-5 | 1[(C)] | 3[(C)] | |
| Chlorine dioxide | 10049-04-4 | 0.1 | 0.3 | |
| Chlorine trifluoride | 7790-91-2 | 0.1[(C)] | 0.4[(C)] | |
| Chloroacetaldehyde | 107-20-0 | 1[(C)] | 3[(C)] | |
| *a*-Chloroacetophenone (phenacyl chloride) | 532-27-4 | 0.05 | 0.3 | |
| Chlorobenzene | 108-90-7 | 75 | 350 | |
| *o*-Chlorobenzylidene malononitrile | 2698-41-1 | 0.05 | 0.4 | |
| Chlorobromomethane | 74-97-5 | 200 | 1050 | |
| 2-Chloro-1,3-butadiene; see *beta*-chloroprene | | | | |
| Chlorodiphenyl (42% chlorine) (PCB) | 53469-21-9 | — | 1 | X |
| Chlorodiphenyl (54% chlorine) (PCB) | 11097-69-1 | — | 0.5 | X |
| 1-Chloro-2,3-epoxypropane; see epichlorohydrin | | | | |
| 2-Chloroethanol; see ethylene chlorohydrin | | | | |
| Chloroethylene; see vinyl chloride | | | | |
| Chloroform (trichloromethane) | 67-66-3 | 50[(C)] | 240[(C)] | |
| *bis*(Chloromethyl) ether; see §1910.1008 | 542-88-1 | — | — | |
| Chloromethyl methyl ether; see §1910.1006 | 107-30-2 | — | — | |
| 1-Chloro-1-nitropropane | 600-25-9 | 20 | 100 | |
| Chloropicrin | 76-06-2 | 0.1 | 0.7 | |
| *beta*-Chloroprene | 126-99-8 | 25 | 90 | X |

***Table Z-1*** Limits for Air Contaminants *(continued)*

| Substance | CAS No.[c] | PEL (ppm)[1,a] | PEL ($mg/m^3$)[1,b] | Skin Designation |
|---|---|---|---|---|
| 2-Chloro-6 (trichloromethyl) pyridine | 1929-82-4 | — | — | |
| Total dust | — | — | 15 | |
| Respirable fraction | — | — | 5 | |
| Chromic acid and chromates (as $CrO_3$) | —[4] | — | —[2] | |
| Chromium (II) compounds (as Cr) | 7440-47-3 | — | 0.5 | |
| Chromium (III) compounds (as Cr) | 7440-47-3 | — | 0.5 | |
| Chromium metal and insoluble salts (as Cr) | 7440-47-3 | — | 1 | |
| Chrysene; see coal tar pitch volatiles | | | | |
| Clopidol | 2971-90-6 | — | — | |
| Total dust | — | — | 15 | |
| Respirable fraction | — | — | 5 | |
| Coal dust (<5% $SiO_2$), respirable fraction | — | — | —[3] | |
| Coal dust (≥5% $SiO_2$), respirable fraction | — | — | —[3] | |
| Coal tar pitch volatiles (benzene soluble fraction), anthracene, BaP, phenanthrene, acridine, chrysene, pyrene | 65966-93-2 | — | 0.2 | |
| Cobalt metal, dust, and fume (as Co) | 7440-48-4 | — | 0.1 | |
| Coke oven emissions; see §1910.1029 | | | | |
| Copper | 7440-50-8 | — | — | |
| Fume (as Cu) | — | — | 0.1 | |
| Dusts and mists (as Cu) | — | — | 1 | |
| Cotton dust;[e] see §1910.1043 | — | — | 1 | |
| Crag herbicide (Sesone) | 136-78-7 | — | — | |
| Total dust | — | — | 15 | |
| Respirable fraction | — | — | 5 | |
| Cresol, all isomers | 1319-77-3 | 5 | 22 | X |
| Crotonaldehyde | 123-73-9 | 2 | 6 | |
| | 4170-30-3 | — | — | |
| Cumene | 98-82-8 | 50 | 245 | X |

***Table Z-1*** Limits for Air Contaminants *(continued)*

| Substance | CAS No.[c] | PEL (ppm)[1,a] | PEL (mg/m$^3$)[1,b] | Skin Designation |
|---|---|---|---|---|
| Cyanides (as CN) | —[4] | — | 5 | X |
| Cyclohexane | 110-82-7 | 300 | 1050 | |
| Cyclohexanol | 108-93-0 | 50 | 200 | |
| Cyclohexanone | 108-94-1 | 50 | 200 | |
| Cyclohexene | 110-83-8 | 300 | 1015 | |
| Cyclopentadiene | 542-92-7 | 75 | 200 | |
| 2,4-D (Dichlorophen-oxyacetic acid) | 94-75-7 | — | 10 | |
| Decaborane | 17702-41-9 | 0.05 | 0.3 | X |
| Demeton (Systox) | 8065-48-3 | — | 0.1 | X |
| Diacetone alcohol (4-hydroxy-4-methyl-2-pentanone) | 123-42-2 | 50 | 240 | |
| 1,2-Diaminoethane; see ethylenediamine | | | | |
| Diazomethane | 334-88-3 | 0.2 | 0.4 | |
| Diborane | 19287-45-7 | 0.1 | 0.1 | |
| 1,2-Dibromo-3-chloropropane (DBCP); see §1910.1044 | 96-12-8 | — | — | |
| 1,2-Dibromoethane; see ethylene dibromide | | | | |
| Dibutyl phosphate | 107-66-4 | 1 | 5 | |
| Dibutyl phthalate | 84-74-2 | — | 5 | |
| *o*-Dichlorobenzene | 95-50-1 | 50[(C)] | 300[(C)] | |
| *p*-Dichlorobenzene | 106-46-7 | 75 | 450 | |
| 3,3'-Dichlorobenzidine; see §1910.1007 | 91-94-1 | — | — | |
| Dichlorodifluoromethane | 75-71-8 | 1000 | 4950 | |
| 1,3-Dichloro-5,5-dimethyl hydantoin— 118-52-5 | — | 0.2 | — | |
| Dichlorodiphenyltri-chloroethane (DDT) | 50-29-3 | — | 1 | X |
| 1,1-Dichloroethane | 75-34-3 | 100 | 400 | |
| 1,2-Dichloroethane; see ethylene dichloride | | | | |
| 1,2-Dichloroethylene | 540-59-0 | 200 | 790 | |
| Dichloroethyl ether | 111-44-4 | 15[(C)] | 90[(C)] | X |
| Dichloromethane; see methylene chloride | | | | |

***Table Z-1*** Limits for Air Contaminants *(continued)*

| Substance | CAS No.[c] | PEL (ppm)[1,a] | PEL (mg/m³)[1,b] | Skin Designation |
|---|---|---|---|---|
| Dichloromonofluoro-methane | 75-43-4 | 1000 | 4200 | |
| 1,1-Dichloro-1-nitroethane | 594-72-9 | 10[(C)] | 60[(C)] | |
| 1,2-Dichloropropane; see propylene dichloride | | | | |
| Dichlorotetrafluoro-ethane | 76-14-2 | 1000 | 7000 | |
| Dichlorvos (DDVP) | 62-73-7 | — | 1 | X |
| Dicyclopentadienyl iron | 102-54-5 | — | — | |
| Total dust | — | — | 15 | |
| Respirable fraction | — | — | 5 | |
| Dieldrin | 60-57-1 | — | 0.25 | X |
| Diethylamine | 109-89-7 | 25 | 75 | |
| 2-Diethylaminoethanol | 100-37-8 | 10 | 50 | X |
| Diethyl ether; see ethyl ether | | | | |
| Difluorodibromomethane | 75-61-6 | 100 | 860 | |
| Diglycidyl ether (DGE) | 2238-07-5 | 0.5[(C)] | 2.8[(C)] | |
| Dihydroxybenzene; see hydroquinone | | | | |
| Diisobutyl ketone | 108-83-8 | 50 | 290 | |
| Diisopropylamine | 108-18-9 | 5 | 20 | X |
| 4-Dimethylaminoazo-benzene; see §1910.1015 | 60-11-7 | — | — | |
| Dimethoxymethane; see methylal | | | | |
| Dimethyl acetamide | 127-19-5 | 10 | 35 | X |
| Dimethylamine | 124-40-3 | 10 | 18 | |
| Dimethylaminobenzene; see xylidine | | | | |
| Dimethylaniline (*N,N*-dimethylaniline) | 121-69-7 | 5 | 25 | X |
| Dimethylbenzene; see xylene | | | | |
| Dimethyl-1,2-dibromo-2,2-dichloroethyl phosphate | 300-76-5 | — | 3 | |
| Dimethylformamide | 68-12-2 | 10 | 30 | X |
| 2,6-Dimethyl-4-heptanone; see diisobutyl ketone | | | | |
| 1,1-Dimethylhydrazine | 57-14-7 | 0.5 | 1 | X |
| Dimethylphthalate | 131-11-3 | — | 5 | |

***Table Z-1*** Limits for Air Contaminants *(continued)*

| Substance | CAS No.[c] | PEL (ppm)[1,a] | PEL ($mg/m^3$)[1,b] | Skin Designation |
|---|---|---|---|---|
| Dimethyl sulfate | 77-78-1 | 1 | 5 | X |
| Dinitrobenzene (all isomers) | — | — | 1 | X |
| *ortho-* | 528-29-0 | — | — | |
| *meta-* | 99-65-0 | — | — | |
| *para-* | 100-25-4 | — | — | |
| Dinitro-*o*-cresol | 534-52-1 | — | 0.2 | X |
| Dinitrotoluene | 25321-14-6 | — | 1.5 | X |
| Dioxane (diethylene dioxide) | 123-91-1 | 100 | 360 | X |
| Diphenyl (biphenyl) | 92-52-4 | 0.2 | 1 | |
| Diphenylmethane diisocyanate; see methylene bisphenyl isocyanate | | | | |
| Dipropylene glycol methyl ether | 34590-94-8 | 100 | 600 | X |
| Di-*sec*-octyl-phthalate (di-(2-ethylhexyl) phthalate) | 117-81-7 | — | 5 | |
| Emery | 12415-34-8 | — | — | |
| Total dust | — | — | 15 | |
| Respirable fraction | — | — | 5 | |
| Endrin | 72-20-8 | — | 0.1 | X |
| Epichlorohydrin | 106-89-8 | 5 | 19 | X |
| EPN | 2104-64-5 | — | 0.5 | X |
| 1,2-Epoxypropane; see propylene oxide | | | | |
| 2,3-Epoxy-1-propanol; see glycidol | | | | |
| Ethanethiol; see ethyl mercaptan | | | | |
| Ethanolamine | 141-43-5 | 3 | 6 | |
| 2-Ethoxyethanol (cellosolve) | 110-80-5 | 200 | 740 | X |
| 2-Ethoxyethyl acetate (cellosolve acetate) | 111-15-9 | 100 | 540 | X |
| Ethyl acetate | 141-78-6 | 400 | 1400 | |
| Ethyl acrylate | 140-88-5 | 25 | 100 | X |
| Ethyl alcohol (ethanol) | 64-17-5 | 1000 | 1900 | |
| Ethylamine | 75-04-7 | 10 | 18 | |
| Ethyl amyl ketone (5-methyl-3-heptanone) | 541-85-5 | 25 | 130 | |

**Table Z-1** Limits for Air Contaminants *(continued)*

| Substance | CAS No.[c] | PEL (ppm)[1,a] | PEL (mg/m³)[1,b] | Skin Designation |
|---|---|---|---|---|
| Ethyl benzene | 100-41-4 | 100 | 435 | |
| Ethyl bromide | 74-96-4 | 200 | 890 | |
| Ethyl butyl ketone (3-heptanone) | 106-35-4 | 50 | 230 | |
| Ethyl chloride | 75-00-3 | 1000 | 2600 | |
| Ethyl ether | 60-29-7 | 400 | 1200 | |
| Ethyl formate | 109-94-4 | 100 | 300 | |
| Ethyl mercaptan | 75-08-1 | 10(C) | 25(C) | |
| Ethyl silicate | 78-10-4 | 100 | 850 | |
| Ethylene chlorohydrin | 107-07-3 | 5 | 16 | X |
| Ethylenediamine | 107-15-3 | 10 | 25 | |
| Ethylene dibromide | 106-93-4 | — | —[2] | |
| Ethylene dichloride (1,2-dichloroethane) | 107-06-2 | — | —[2] | |
| Ethylene glycol dinitrate | 628-96-6 | 0.2(C) | 1(C) | X |
| Ethylene glycol methyl acetate; see methyl cellosolve acetate | | | | |
| Ethyleneimine; see §1910.1012 | 151-56-4 | — | — | |
| Ethylene oxide; see §1910.1047 | 75-21-8 | — | — | |
| Ethylidene chloride; see 1,1-dichlorethane | | | | |
| *N*-Ethylmorpholine | 100-74-3 | 20 | 94 | X |
| Ferbam | 14484-64-1 | — | — | |
| Total dust | — | — | 15 | |
| Ferrovanadium dust | 12604-58-9 | — | 1 | |
| Fluorides (as F) | —[4] | — | 2.5 | |
| Fluorine | 7782-41-4 | 0.1 | 0.2 | |
| Fluorotrichloromethane (trichlorofluoromethane) | 75-69-4 | 1000 | 5600 | |
| Formaldehyde; see §1910.1048 | 50-00-0 | — | — | |
| Formic acid | 64-18-6 | 5 | 9 | |
| Furfural | 98-01-1 | 5 | 20 | X |
| Furfuryl alcohol | 98-00-0 | 50 | 200 | |
| Grain dust (oat, wheat barley) | — | — | 10 | |

***Table Z-1*** Limits for Air Contaminants *(continued)*

| Substance | CAS No.[c] | PEL (ppm)[1,a] | PEL (mg/m³)[1,b] | Skin Designation |
|---|---|---|---|---|
| Glycerin (mist) | 56-81-5 | — | — | |
| Total dust | — | — | 15 | |
| Respirable fraction | — | — | 5 | |
| Glycidol | 556-52-5 | 50 | 150 | |
| Glycol monoethyl ether; see 2-ethoxyethanol | | | | |
| Graphite, natural respirable dust | 7782-42-5 | — | —[3] | |
| Graphite, synthetic | — | — | — | |
| Total dust | — | — | 15 | |
| Respirable fraction | — | — | 5 | |
| Guthion; see azinphos methyl | | | | |
| Gypsum | 13397-24-5 | — | — | |
| Total dust | — | — | 15 | |
| Respirable fraction | — | — | 5 | |
| Hafnium | 7440-58-6 | — | 0.5 | |
| Heptachlor | 76-44-8 | — | 0.5 | X |
| Heptane (*n*-Heptane) | 142-82-5 | 500 | 2000 | |
| Hexachloroethane | 67-72-1 | 1 | 10 | X |
| Hexachloronaphthalene | 1335-87-1 | — | 0.2 | X |
| *n*-Hexane | 110-54-3 | 500 | 1800 | |
| 2-Hexanone (methyl *n*-butyl ketone) | 591-78-6 | 100 | 410 | |
| Hexone (methyl isobutyl ketone) | 108-10-1 | 100 | 410 | |
| *sec*-Hexyl acetate | 108-84-9 | 50 | 300 | |
| Hydrazine | 302-01-2 | 1 | 1.3 | X |
| Hydrogen bromide | 10035-10-6 | 3 | 10 | |
| Hydrogen chloride | 7647-01-0 | 5[(C)] | 7[(C)] | |
| Hydrogen cyanide | 74-90-8 | 10 | 11 | X |
| Hydrogen fluoride (as F) | 7664-39-3 | — | —[2] | |
| Hydrogen peroxide | 7722-84-1 | 1 | 1.4 | |
| Hydrogen selenide (as Se) | 7783-07-5 | 0.05 | 0.2 | |
| Hydrogen sulfide | 7783-06-4 | — | —[2] | |
| Hydroquinone | 123-31-9 | — | —[2] | |
| Iodine | 7553-56-2 | 0.1[(C)] | 1[(C)] | |
| Iron oxide fume | 1309-37-1 | — | 10 | |

**Table Z-1** Limits for Air Contaminants *(continued)*

| Substance | CAS No.[c] | PEL (ppm)[1,a] | PEL (mg/m³)[1,b] | Skin Designation |
|---|---|---|---|---|
| Isomyl acetate | 123-92-2 | 100 | 525 | |
| Isomyl alcohol (primary and secondary) | 123-51-3 | 100 | 360 | |
| Isobutyl acetate | 110-19-0 | 150 | 700 | |
| Isobutyl alcohol | 78-83-1 | 100 | 300 | |
| Isophorone | 78-59-1 | 25 | 140 | |
| Isopropyl acetate | 108-21-4 | 250 | 950 | |
| Isopropyl alcohol | 67-63-0 | 400 | 980 | |
| Isopropylamine | 75-31-0 | 5 | 12 | |
| Isopropyl ether | 108-20-3 | 500 | 2100 | |
| Isopropyl glycidyl ether (IGE) | 4016-14-2 | 50 | 240 | |
| Kaolin | 1332-58-7 | — | — | |
| Total dust | — | — | 15 | |
| Respirable fraction | — | — | 5 | |
| Ketene | 463-51-4 | 0.5 | 0.9 | |
| Lead inorganic (as Pb); see §1910.1025 | 7439-92-1 | — | — | |
| Limestone | 1317-65-3 | — | — | |
| Total dust | — | — | 15 | |
| Respirable fraction | — | — | 5 | |
| Lindane | 58-89-9 | — | 0.5 | X |
| Lithium hydride | 7580-67-8 | — | 0.025 | |
| L.P.G. (liquified petroleum gas) | 68476-85-7 | 1000 | 1800 | |
| Magnesite | 546-93-0 | — | — | |
| Total dust | — | — | 15 | |
| Respirable fraction | — | — | 5 | |
| Magnesium oxide fume | 1309-48-4 | — | — | |
| Total particulate | — | — | 15 | |
| Malathion | 121-75-5 | — | — | |
| Total dust | — | — | 15 | X |
| Maleic anhydride | 108-31-6 | 0.25 | 1 | |
| Manganese compounds (as Mn) | 7439-96-5 | — | 5[(C)] | |
| Manganese fume (as Mn) | 7439-96-5 | — | 5[(C)] | |
| Marble | 1317-65-3 | | | |
| Total dust | — | — | 15 | |
| Respirable fraction | — | — | 5 | |

***Table Z-1*** Limits for Air Contaminants *(continued)*

| Substance | CAS No.[c] | PEL (ppm)[1,a] | PEL (mg/m³)[1,b] | Skin Designation |
|---|---|---|---|---|
| Mercury (aryl and inorganic) (as Hg) | 7439-97-6 | — | —[2] | |
| Mercury (organo) alkyl compounds (as Hg) | 7439-97-6 | — | —[2] | |
| Mercury (vapor) (as Hg) | 7439-97-6 | — | —[2] | |
| Mesityl oxide | 141-79-7 | 25 | 100 | |
| Methanethiol; see methyl mercaptan | | | | |
| Methoxychlor | 72-43-5 | — | — | |
| Total dust | — | — | 15 | |
| 2-Methoxyethanol (methyl cellosolve) | 109-86-4 | 25 | 80 | X |
| 2-Methoxyethyl acetate (methyl cellosolve acetate) | 110-49-6 | 25 | 120 | X |
| Methyl acetate | 79-20-9 | 200 | 610 | |
| Methyl acetylene (propyne) | 74-99-7 | 1000 | 1650 | |
| Methyl acetylene propadiene mixture (MAPP) | — | 1000 | 1800 | |
| Methyl acrylate | 96-33-3 | 10 | 35 | X |
| Methylal (dimethoxy-methane) | 109-87-5 | 1000 | 3100 | |
| Methyl alcohol | 67-56-1 | 200 | 260 | |
| Methylamine | 74-89-5 | 10 | 12 | |
| Methyl amyl alcohol; see methyl isobutyl carbinol | | | | |
| Methyl *n*-amyl ketone | 110-43-0 | 100 | 465 | |
| Methyl bromide | 74-83-9 | 20(C) | 80(C) | X |
| Methyl butyl ketone; see 2-hexanone | | | | |
| Methyl cellosolve; see 2-methoxyethanol | | | | |
| Methyl cellosolve acetate; see 2-methoxyethyl acetate | | | | |
| Methyl chloride | 74-87-3 | — | —[2] | |
| Methyl chloroform (1,1,1-trichloroethane) | 71-55-6 | 350 | 1900 | |
| Methylcyclohexane | 108-87-2 | 500 | 2000 | |
| Methylcyclohexanol | 25639-42-3 | 100 | 470 | |
| *o*-Methylcyclohexanone | 583-60-8 | 100 | 460 | X |

**Table Z-1** Limits for Air Contaminants *(continued)*

| Substance | CAS No.[c] | PEL (ppm)[1,a] | PEL (mg/m³)[1,b] | Skin Designation |
|---|---|---|---|---|
| Methylene bisphenyl isocyanate (MDI) | 101-68-8 | 0.02[(C)] | 0.2[(C)] | |
| Methylene chloride | 75-09-2 | — | —[2] | |
| Methyl ethyl ketone (MEK); see 2-butanone | | | | |
| Methyl formate | 107-31-3 | 100 | 250 | |
| Methyl hydrazine (monomethyl hydrazine) | 60-34-4 | 0.2[(C)] | 0.35[(C)] | X |
| Methyl iodide | 74-88-4 | 5 | 28 | X |
| Methyl isoamyl ketone | 110-12-3 | 100 | 475 | |
| Methyl isobutyl carbinol | 108-11-2 | 25 | 100 | X |
| Methyl isobutyl ketone; see hexone | | | | |
| Methyl isocyanate | 624-83-9 | 0.02 | 0.05 | |
| Methyl mercaptan | 74-93-1 | 10[(C)] | 20[(C)] | |
| Methyl methacrylate | 80-62-6 | 100 | 410 | |
| Methal propyl ketone; see 2-pentanone | | | | |
| *alpha*-Methyl styrene | 98-83-9 | 100[(C)] | 480[(C)] | |
| Methylene bisphenyl isocyanate (MDI) | 101-68-8 | 0.02[(C)] | 0.2[(C)] | |
| Mica; see silicates | | | | |
| Molybdenum (as Mo) | 7439-98-7 | — | — | |
| Soluble compounds | — | — | 5 | |
| Insoluble compounds | — | — | — | |
| Total dust | — | — | 15 | |
| Monomethyl aniline | 100-61-8 | 2 | 9 | X |
| Monomethyl hydrazine; see methyl hydrazine | | | | |
| Morpholine | 110-91-8 | 20 | 70 | X |
| Naphtha (coal tar) | 8030-30-6 | 100 | 400 | |
| Naphthalene | 91-20-3 | 10 | 50 | |
| *alpha*-Naphthylamine; see §1910.1004 | 134-32-7 | — | — | |
| *beta*-Naphthylamine; see §1910.1009 | 91-59-8 | — | — | |
| Nickel carbonyl (as Ni) | 13463-39-3 | 0.001 | 0.007 | |
| Nickel, metal and insoluble compounds (as Ni) | 7440-02-0 | — | 1 | |
| Nickel, soluble compounds (as Ni) | 7440-02-0 | — | 1 | |

***Table Z-1*** Limits for Air Contaminants *(continued)*

| Substance | CAS No.[c] | PEL (ppm)[1,a] | PEL (mg/m³)[1,b] | Skin Designation |
|---|---|---|---|---|
| Nicotine | 54-11-5 | — | 0.5 | |
| Nitric acid | 7697-37-2 | 2 | 5 | |
| Nitric oxide | 10102-43-9 | 25 | 30 | |
| *p*-Nitroaniline | 100-01-6 | 1 | 6 | X |
| Nitrobenzene | 98-95-3 | 1 | 5 | X |
| *p*-Nitrochlorobenzene | 100-00-5 | — | 1 | X |
| 4-Nitrodiphenyl; see §1910.1003 | 92-93-3 | | | |
| Nitroethane | 79-24-3 | 100 | 310 | |
| Nitrogen dioxide | 10102-44-0 | 5[(C)] | 9[(C)] | |
| Nitrogen trifluoride | 7783-54-2 | 10 | 29 | |
| Nitroglycerin | 55-63-0 | 0.2[(C)] | 2[(C)] | X |
| Nitromethane | 75-52-5 | 100 | 250 | |
| 1-Nitropropane | 108-03-2 | 25 | 90 | |
| 2-Nitropropane | 79-46-9 | 25 | 90 | |
| *N*-Nitrosodimethylamine; see §1910.1016 | | | | |
| Nitrotoluene (all isomers) | — | 5 | 30 | X |
| *o*-isomer | 88-72-2 | — | — | |
| *m*-isomer | 99-08-1 | — | — | |
| *p*-isomer | 99-99-0 | — | — | |
| Nitrotrichloromethane; see chloropicrin | | | | |
| Octachloronaphthalene | 2234-13-1 | — | 0.1 | X |
| Octane | 111-65-9 | 500 | 2350 | |
| Oil mist, mineral | 8012-95-1 | — | 5 | |
| Osmium tetroxide (as Os) | 20816-12-0 | — | 0.002 | |
| Oxalic acid | 144-62-7 | — | 1 | |
| Oxygen difluoride | 7783-41-7 | 0.05 | 0.1 | |
| Ozone | 10028-15-6 | 0.1 | 0.2 | |
| Paraquat, respirable dust | 4685-14-7 | — | 0.5 | X |
| | 1910-42-5 | — | — | |
| | 2074-50-2 | — | — | |
| Parathion | 56-38-2 | — | 0.1 | X |
| Particulates not otherwise regulated (PNOR)[f] | — | — | — | |
| Total dust | — | — | 15 | |
| Respirable fraction | — | — | 5 | |

**Table Z-1** Limits for Air Contaminants *(continued)*

| Substance | CAS No.[c] | PEL (ppm)[1,a] | PEL (mg/m³)[1,b] | Skin Designation |
|---|---|---|---|---|
| PCB; see chlorodiphenyl (42% and 54% chlorine) | | | | |
| Pentaborane | 19624-22-7 | 0.005 | 0.01 | |
| Pentachloronaphthalene | 1321-64-8 | — | 0.5 | X |
| Pentachlorophenol | 87-86-5 | — | 0.5 | X |
| Pentaerythritol | 115-77-5 | | | |
| Total dust | — | — | 15 | |
| Respirable fraction | — | — | 5 | |
| Pentane | 109-66-0 | 1000 | 2950 | |
| 2-Pentanone (methyl propyl ketone) | 107-87-9 | 200 | 700 | |
| Perchloroethylene (tetrachloroethylene) | 127-18-4 | — | —[2] | |
| Perchloromethyl mercaptan | 594-42-3 | 0.1 | 0.8 | |
| Perchloryl fluoride | 7616-94-6 | 3 | 13.5 | |
| Petroleum distillates (naphtha) (rubber solvent) | — | 500 | 2000 | |
| Phenol | 108-95-2 | 5 | 19 | X |
| *p*-Phenylene diamine | 106-50-3 | — | 0.1 | X |
| Phenyl ether, vapor | 101-84-8 | 1 | 7 | |
| Phenyl ether-biphenyl mixture, vapor | — | 1 | 7 | |
| Phenylethylene; see styrene | | | | |
| Phenyl glycidyl ether (PGE) | 122-60-1 | 10 | 60 | |
| Phenylhydrazine | 100-63-0 | 5 | 22 | X |
| Phosdrin (mevinphos) | 7786-34-7 | — | 0.1 | X |
| Phosgene (carbonyl chloride) | 75-44-5 | 0.1 | 0.4 | |
| Phosphine | 7803-51-2 | 0.3 | 0.4 | |
| Phosphoric acid | 7664-38-2 | — | 1 | |
| Phosphorus (yellow) | 7723-14-0 | — | 0.1 | |
| Phosphorus pentachloride | 10026-13-8 | — | 1 | |
| Phosphorus pentasulfide | 1314-80-3 | — | 1 | |
| Phosphorus trichloride | 7719-12-2 | 0.5 | 3 | |
| Phthalic anhydride | 85-44-9 | 2 | 12 | |
| Picloram | 1918-02-1 | — | — | |
| Total dust | — | — | 15 | |
| Respirable fraction | — | — | 5 | |
| Picric acid | 88-89-1 | — | 0.1 | X |

***Table Z-1*** Limits for Air Contaminants *(continued)*

| Substance | CAS No.[c] | PEL (ppm)[1,a] | PEL (mg/m$^3$)[1,b] | Skin Designation |
|---|---|---|---|---|
| Pindone (2-pivalyl-1,3-indandione) | 83-26-1 | — | 0.1 | |
| Plaster of Paris | 26499-65-0 | — | — | |
| Total dust | — | — | 15 | |
| Respirable fraction | — | — | 5 | |
| Platinum (as Pt) | 7440-06-4 | — | — | |
| Metal | — | — | — | |
| Soluble salts | — | — | 0.002 | |
| Portland cement | 65997-15-1 | — | — | |
| Total dust | — | — | 15 | |
| Respirable fraction | — | — | 5 | |
| Propane | 74-98-6 | 1000 | 1800 | |
| *beta*-Propriolactone; see §1910.1013 | 57-57-8 | | | |
| *n*-Propyl acetate | 109-60-4 | 200 | 840 | |
| *n*-Propyl alcohol | 71-23-8 | 200 | 500 | |
| *n*-Propyl nitrate | 627-13-4 | 25 | 110 | |
| Propylene dichloride | 78-87-5 | 75 | 350 | |
| Propylene imine | 75-55-8 | 2 | 5 | X |
| Propylene oxide | 75-56-9 | 100 | 240 | |
| Propyne; see methyl acetylene | | | | |
| Pyrethrum | 8003-34-7 | — | 5 | |
| Pyridine | 110-86-1 | 5 | 15 | |
| Quinone | 106-51-4 | 0.1 | 0.4 | |
| RDX; see cyclonite | | | | |
| Rhodium (as Rh), metal fume and insoluble compounds | 7440-16-6 | — | 0.1 | |
| Rhodium (as Rh), soluble compounds | 7440-16-6 | — | 0.001 | |
| Ronnel | 299-84-3 | — | 15 | |
| Rotenone | 83-79-4 | — | 5 | |
| Rouge | — | — | — | |
| Total dust | — | — | 15 | |
| Respirable fraction | — | — | 5 | |
| Selenium compounds (as Se) | 7782-49-2 | — | 0.2 | |
| Selenium hexafluoride (as Se) | 7783-79-1 | 0.05 | 0.4 | |

***Table Z-1*** Limits for Air Contaminants *(continued)*

| Substance | CAS No.[c] | PEL (ppm)[1,a] | PEL (mg/m³)[1,b] | Skin Designation |
|---|---|---|---|---|
| Silica, amorphous, precipitated and gel | 112926-00-8 | — | —[3] | |
| Silica, amorphous, diatomaceous earth, containing less than 1% crystalline silica | 1790-53-2 | — | —[3] | |
| Silica, crystalline cristobalite, respirable dust | 14464-46-1 | — | —[3] | |
| Silica, crystalline quartz, respirable dust | 14808-60-7 | — | —[3] | |
| Silica, crystalline tripoli (as quartz), respirable dust | 1317-95-9 | — | —[3] | |
| Silica, crystalline tridymite, respirable dust | 15468-32-3 | — | —[3] | |
| Silica, fused, respirable dust | 60676-86-0 | — | —[3] | |
| Silicates (less than 1% crystalline silica) | | | | |
| Mica (respirable dust) | 12001-26-2 | — | —[3] | |
| Soapstone, total dust | — | — | —[3] | |
| Soapstone, respirable dust | — | — | —[3] | |
| Talc (containing asbestos): use asbestos limit; see §1910.1001 | — | — | —[3] | |
| Talc (containing no asbestos), respirable dust | 14807-96-6 | — | —[3] | |
| Tremolite, asbestiform; see §1910.1001 | | | | |
| Silicon | 7440-21-3 | — | — | |
| Total dust | — | — | 15 | |
| Respirable fraction | — | — | 5 | |
| Silicon carbide | 409-21-2 | | | |
| Total dust | — | — | 15 | |
| Respirable fraction | — | — | 5 | |
| Silver, metal and soluble compounds (as Ag) | 7440-22-4 | — | 0.01 | |
| Soapstone; see silicates | | | | |
| Sodium fluoroacetate | 62-74-8 | — | 0.05 | X |
| Sodium hydroxide | 1310-73-2 | — | 2 | |
| Starch | 9005-25-8 | — | — | |
| Total dust | — | — | 15 | |
| Respirable fraction | — | — | 5 | |

***Table Z-1*** Limits for Air Contaminants *(continued)*

| Substance | CAS No.[c] | PEL (ppm)[1,a] | PEL (mg/m³)[1,b] | Skin Designation |
|---|---|---|---|---|
| Stibine | 7803-52-3 | 0.1 | 0.5 | |
| Stoddard solvent | 8052-41-3 | 500 | 2900 | |
| Strychnine | 57-24-9 | — | 0.15 | |
| Styrene | 100-42-5 | —[2] | — | |
| Sucrose | 57-50-1 | — | — | |
| Total dust | — | — | 15 | |
| Respirable fraction | — | — | 5 | |
| Sulfur dioxide | 7446-09-5 | 5 | 13 | |
| Sulfur hexafluoride | 2551-62-4 | 1000 | 6000 | |
| Sulfuric acid | 7664-93-9 | — | 1 | |
| Sulfur monochloride | 10025-67-9 | 1 | 6 | |
| Sulfur pentafluoride | 5714-22-7 | 0.025 | 0.25 | |
| Sulfuryl fluoride | 2699-79-8 | 5 | 20 | |
| Systox; see demeton | | | | |
| 2,4,5-T (2,4,5-trichloro-phenoxyacetic acid) | 93-76-5 | — | 10 | |
| Talc; see silicates | | | | |
| Tantalum, metal and oxide dust | 7440-25-7 | — | 5 | |
| TEDP (Sulfotep) | 3689-24-5 | — | 0.2 | X |
| Tellurium and compounds (as Te) | 13494-80-9 | — | 0.1 | |
| Tellurium hexafluoride (as Te) | 7783-80-4 | 0.02 | 0.2 | |
| Temephos | 3383-96-8 | — | — | |
| Total dust | — | — | 15 | |
| Respirable fraction | — | — | 5 | |
| TEPP (tetraethyl pyrophosphaate) | 107-49-3 | — | 0.05 | X |
| Terphenylis | 26140-60-3 | 1[(C)] | 9[(C)] | |
| 1,1,1,2-Tetrachloro-2,2-difluoroethane | 76-11-9 | 500 | 4170 | |
| 1,1,2,2-Tetrachloro-1,2-difluoroethane | 76-12-0 | 500 | 4170 | |
| 1,1,2,2-Tetrachloroethane | 79-34-5 | 5 | 35 | X |
| Tetrachoroethylene; see perchloroethylene | | | | |
| Tetrachloromethane; see carbon tetrachloride | | | | |
| Tetrachloronaphthalene | 1335-88-2 | — | 2 | X |

**Table Z-1** Limits for Air Contaminants *(continued)*

| Substance | CAS No.[c] | PEL (ppm)[1,a] | PEL (mg/m³)[1,b] | Skin Designation |
|---|---|---|---|---|
| Tetraethyl lead (as Pb) | 78-00-2 | — | 0.075 | X |
| Tetrahydrofuran | 109-99-9 | 200 | 590 | |
| Tetramethyl lead (as Pb) | 75-74-1 | — | 0.075 | X |
| Tetramethyl succinonitrile | 3333-52-6 | 0.5 | 3 | X |
| Tetranitromethane | 509-14-8 | 1 | 8 | |
| Tetryl (2,4,6-trinitrophenyl-methyl-nitramine) | 479-45-8 | — | 1.5 | X |
| Thallium, soluble compounds (as Tl) | 7440-28-0 | — | 0.1 | X |
| 4,4'-Thiobis (6-*tert*, *butyl-m*-cresol) | 96-69-5 | — | — | |
| Total dust | — | — | 15 | |
| Respirable fraction | — | — | 5 | |
| Thiram | 137-26-8 | — | 5 | |
| Tin, inorganic compounds (except oxides) (as Sn) | 7440-31-5 | — | 2 | |
| Tin, organic compounds (as Sn) | 7440-31-5 | — | 0.1 | |
| Titanium dioxide | 13463-67-7 | — | — | |
| Total dust | — | — | 15 | |
| Toluene | 108-88-3 | — | —[2] | |
| Toluene-2,4-diisocyanate (TDI) | 584-84-9 | 0.02[(C)] | 0.14[(C)] | |
| *o*-Toluidine | 95-53-4 | 5 | 22 | X |
| Toxaphene; see chlorinated camphene | | | | |
| Tremolite; see silicates | | | | |
| Tributyl phosphate | 126-73-8 | — | 5 | |
| 1,1,1-Trichloroethane; see methyl chloroform | | | | |
| 1,1,2-Trichloroethane | 79-00-5 | 10 | 45 | X |
| Trichloroethylene | 79-01-6 | — | —[2] | |
| Trichloromethane; see chloroform | | | | |
| Trichloronaphthalene | 1321-65-9 | — | 5 | X |
| 1,2,3-Trichloropropane | 96-18-4 | 50 | 300 | |
| 1,1,2-Trichloro-1,2,2-trifluoroethane | 76-13-1 | 1000 | 7600 | |
| Triethylamine | 121-44-8 | 25 | 100 | |

***Table Z-2*** Toxic and Hazardous Substances

| Substance | 8-hr TWA | Acceptable Ceiling Concentration | Acceptable Maximum Peak above Acceptable Ceiling Concentration for 8-hr Shift | |
|---|---|---|---|---|
| | | | Concentration | Maximum Duration |
| Benzene (a) (Z37.40-1969) | 10 ppm | 25 ppm | 50 ppm | 10 min. |
| Beryllium and beryllium compounds (Z37.29-1970) | 2 μg/m$^3$ | 5 μg/m$^3$ | 25 μg/m$^3$ | 30 min. |
| Cadmium fume (b) (Z37.5-1970) | 0.1 mg/m$^3$ | 0.3 mg/m$^3$ | — | — |
| Cadmium dust (b) (Z37.5-1970) | 0.2 mg/m$^3$ | 0.6 mg/m$^3$ | — | — |
| Carbon disulfide (Z37.3-1968) | 20 ppm | 30 ppm | 100 ppm | 30 min. |
| Carbon tetrachloride (Z37.17-1967) | 10 ppm | 25 ppm | 200 ppm | 5 min. in any 4 hr. |
| Chromic acid and chromates (Z37-7-1971) | — | 1 mg/10 m$^3$ | — | — |
| Ethylene dibromide (Z37.31-1970) | 20 ppm | 30 ppm | 50 ppm | 5 min. |
| Ethylene dichloride (Z37.21-1969) | 50 ppm | 100 ppm | 200 ppm | 5 min. any 3 hr. |
| Fluoride as dust (Z37.28-1969) | 2.5 mg/m$^3$ | — | — | — |
| Formaldehyde; see §1910.1048 | | | | |
| Hydrogen fluoride (Z37.28-1969) | 3 ppm | — | — | — |
| Hydrogen sulfide (Z37.2-1966) | — | 20 ppm | 50 ppm | 10 min. once only if no other measurable exposure occurs |
| Mercury (Z37.8-1971) | 1 mg/10 m$^3$ | — | — | — |
| Methylene chloride (Z37.18-1969); see §1910.1052 | | | | |

***Table Z-2*** Toxic and Hazardous Substances *(continued)*

| Substance | 8-hr TWA | Acceptable Ceiling Concentration | Acceptable Maximum Peak above Acceptable Ceiling Concentration for 8-hr Shift | |
|---|---|---|---|---|
| | | | Concentration | Maximum Duration |
| Organo (alkyl) mercury (Z37.30-1969) | 0.01 mg/m$^3$ | 0.04 mg/m$^3$ | — | — |
| Styrene (Z37.15-1969) | 100 ppm | 200 ppm | 600 ppm | 5 min. in any 3 hr. |
| Tetrachloroethylene (Z37.22-1967) | 100 ppm | 200 ppm | 300 ppm | 5 min. in any 3 hr. |
| Toluene (Z37.12-1967) | 200 ppm | 300 ppm | 500 ppm | 10 min. |
| Trichloroethylene (Z37.19-1967) | 100 ppm | 200 ppm | 300 ppm | 5 min. in any 2 hr. |

# appendix G

# Workplace violence guidelines for night retail establishments

*Important note:* This is a working draft document (dated June 28, 1996) for discussion and comment only. Do not cite or quote as though it is a final document. The document does not represent OSHA's final position with respect to the matters discussed therein. It will be revised based upon comments received from stakeholders. The final document will be advisory in nature and informational in content for employers seeking to provide workplace violence prevention strategies. The issuance of these guidelines is not intended as an enhancement of the General Duty Clause nor does it reflect any change in OSHA enforcement policy.

## Introduction

Workplace violence has emerged as a critical safety and health hazard. According to the Bureau of Labor Statistics (BLS), homicide is the second leading cause of death to American workers, claiming the lives of 1071 workers in 1994 and accounting for 16% of the 6588 fatal work injuries in the United States. Research indicates that measures can be taken to reduce the risks of workplace violence to American workers. OSHA's violence prevention guidelines provide the agency's recommendations for reducing workplace violence developed following a careful review of workplace violence studies, public and private violence prevention programs, and consultation with and input from stakeholders.

OSHA encourages employers to establish violence prevention programs and to track their progress in reducing work-related assaults. Although not every incident can be prevented, many can, and the severity of injuries sustained by employees can be reduced. Adopting practical measures such as those outlined here can significantly reduce this serious threat to worker safety.

## *OSHA's commitment*

The publication and distribution of these guidelines are OSHA's first step in assisting the night retail industry in preventing workplace violence. OSHA plans to conduct a coordinated effort consisting of research, information, training, cooperative programs, and appropriate enforcement to accomplish this goal. The guidelines are not a new standard or regulation. They are advisory in nature, informational in content, and intended for use by employers in providing a safe and healthful workplace through effective violence prevention programs, adapted to the needs and resources of each place of employment.

## *Extent of problem*

### *Fatalities*

Workers in retail establishments face an above-average risk for violence. Of the 1071 deaths due to workplace violence in 1994, as in prior years, half occurred during robberies of small retail establishments, including grocery or convenience stores, restaurants and bars, liquor stores, fast-food restaurants, and gas stations. During 1980–89, there were 2787 homicides in small retail establishments — the highest number of occupational homicides. Shootings accounted for four fifths of fatal workplace assaults, with robbery being the motive in 75% of the homicides.

### *Who are the victims?*

At 82%, men account for the majority of all homicide victims. Homicide, however, is the leading way in which women are killed in the workplace, accounting for 40% of such deaths. Women accounted for 53% of the homicides reported in the retail trades. African-Americans, Asian-Americans, and other minority races comprise an eighth of the work force but constitute a fourth of all workplace homicide victims. This higher homicide risk is due, in part, to their constituting a disproportionate share of the work force in these occupations.

### *Nonfatal assaults*

Workplace homicides are only part of the problem. According to BLS statistics, about 21,300 workers were injured in nonfatal assaults in the workplace in 1993, with women as victims in 56% of these assaults (refer to BLS

*Survey of Occupational Injuries and Illnesses,* 1992/1993). The Department of Justice's *National Crime Victimization Survey* also reported that, between 1987 and 1992, approximately 1 million persons annually were assaulted at work. These data include four categories: 615,160 simple assaults, 264,174 aggravated assaults, 79,109 robberies, and 13,068 rapes. Such intentional injuries to workers occur much more frequently than occupational homicides, but efforts to prevent homicides also may reduce the incidences of these nonfatal assaults.

## *Risk factors/summary of research*

In 1992, the Centers for Disease Control (CDC) declared workplace homicide a serious public health problem. The National Institute for Occupational Safety and Health (NIOSH) identified six factors that increase the risks of homicide in the workplace. All of these factors are usually present in night retail establishments:

- Exchange of money with the public
- Working alone or in small numbers
- Working late night or early morning hours
- Working in high-crime areas
- Guarding valuable property or possessions
- Working in community settings

Several studies have examined risk factors for robbery. A groundbreaking experiment conducted at the request of the Southland Corporation (owner of the 7-Eleven convenience store chain), for example, enlisted the aid of ex-convicts to determine which stores were most "attractive" to robbers. The study confirmed that robberies are not randomly distributed, but robbers are selective in their targets. Stores most vulnerable had large amounts of cash on hand, an obstructed view of counters with inattentive clerks, poor outdoor lighting, and easy escape routes.

Studies conducted by the State of Florida's Attorney General's office and the State of Virginia's Crime Commission have both corroborated these results and added an additional risk factor: clerks working alone at night. According to these studies, the top five "attractiveness" features for choosing a store to rob are remote area, only one clerk, no customers, easy access/getaway, and lots of money on hand.

The five most "unattractive" (deterrent) features were many customers, heavy traffic in store front, two clerks, a visible back room where another employee might be, and a male clerk. It was determined that the primary deterrent was having two clerks on duty. Other deterrents significantly reducing the potential for robbery included: security cameras, time-release safes, a location near other 24-hour businesses, and a midnight store closing time. Retail robberies occur in the late evening and early morning hours more often than during daylight hours. Various studies have found robbery rates to be 65% from 11:00 p.m. to 6:00 a.m., 65% from 9:00 p.m. to 3:00 a.m., and 69% for the time frame 9:00 p.m. to 5:00 a.m. Generally, a lone clerk is working at night when an attack occurs. According to studies of robberies in Gainesville, FL, from 1981 to 1986, 92% occurred when one clerk was working. In 85% of the cases, the clerk was completely alone; no customers were present in the store.

### *Preventive strategies*

"Situational crime prevention" both reduces opportunities for crime and increases its risks. The situational approach is based on the presumption that perpetrators of crime "rationally" select their targets and that the chance of victimization can be reduced by: increasing the effort (target hardening, controlling access, deflecting offenders, and controlling facilitators), increasing the risks (entry/exit screening, formal surveillance, surveillance by employees and others), and reducing the rewards (removing the target, identifying property, removing inducements, and setting rules). Physical and behavioral changes at a site can substantially reduce robberies. A test group of stores eliminated or reduced the risk factors and subsequently experienced a 30% drop in robberies compared to a control group, thus supporting the hypotheses that robbers select their targets and that physical change to the establishment can reduce the incidence or change the location of robberies.

The target-hardening efforts, including a basic robbery deterrence package, were implemented in 7-Eleven stores nationwide in 1976. The 7-Eleven store robberies decreased by 65% from 1976 to 1986 and have held at a 50% reduction since that time. The National Association of Convenience Stores (NACS) adopted the program for use nationwide in 1987. The rationale of the NACS program is to make the target less attractive by reducing the cash, maximizing the take/risk ratio, and training employees. The program includes the following:

- Clearing windows for increased visibility; improving lighting
- Maintaining low cash in register
- Installing time-controlled drop safes
- Posting signs about low cash
- Altering escape routes
- Training employees in not resisting

Employing two clerks is a form of target hardening because it may make a robbery more difficult to complete and, therefore, more unsuitable. Having two clerks may also increase the likelihood that they will be effective witnesses in a police investigation. Since the enactment of the Gainesville ordinance requiring two clerks, robbery was reduced by 92% between the hours of 8:00 p.m. and 4:00 a.m.

### *Overview of guidelines*

In January 1989, OSHA published voluntary, generic, safety and health program management guidelines for all employers to use as a foundation for their safety and health programs, including a workplace violence prevention program. OSHA's violence prevention guidelines build on the 1989 generic guidelines by identifying common risk factors and describing some feasible solutions. Although not exhaustive, the new workplace violence guidelines include policy recommendations and practical corrective methods to help prevent and mitigate the effects of workplace violence. The goal is to eliminate or reduce worker exposure to conditions that lead to death or injury from violence by implementing effective security devices and administrative work practices, among other control measures. These guidelines are intended to cover a broad spectrum of workers in retail trades who provide services during evening and night hours. They are particularly appropriate for workers in convenience stores, liquor stores, and gasoline stations with grocery stores providing services late at night. It is anticipated, however, that other types of establishments — such as drug stores, grocery stores, supermarkets, and eating and drinking establishments — may find these recommendations helpful.

## *Violence prevention program elements*

There are four main components to any effective safety and health program that also apply to preventing workplace violence:

- Management commitment and employee involvement
- Worksite analysis
- Hazard prevention and control
- Safety and health training

### *Management commitment and employee involvement*

Management commitment and employee involvement are complementary and essential elements of an effective safety and health program. To ensure an effective program, management and front-line employees must work together, perhaps through a team or committee approach. If employers opt for this strategy, they must be careful to comply with the applicable provisions of the National Labor Relations Act which prohibits unfair labor practices. Management commitment, including the endorsement and visible involvement of top management, provides the motivation and resources to deal effectively with workplace violence, and should include the following:

- Demonstrated organizational concern for employee emotional and physical safety and health
- Equal commitment to worker safety and health and customer safety

- Assigned responsibility for the various aspects of the workplace violence prevention program to ensure that all managers, supervisors, and employees understand their obligations
- Appropriate allocation of authority and resources to all responsible parties
- A system of accountability for involved managers, supervisors, and employees
- A comprehensive program of medical and psychological counseling and debriefing for employees experiencing or witnessing assaults and other violent incidents
- Commitment to support and implement appropriate recommendations from safety and health committees

Employee involvement and feedback enable workers to develop and express their own commitment to safety and health and provide useful information to design, implement, and evaluate the program. Employee involvement should include the following:

- Understanding and complying with the workplace violence prevention program and other safety and security measures
- Participation in an employee complaint or suggestion procedure covering safety and security concerns
- Prompt and accurate reporting of violent incidents
- Participation on safety and health committees or teams that receive reports of violent incidents or security problems, make facility inspections, and respond with recommendations for corrective strategies
- Taking part in a continuing education program that covers techniques to recognize escalating agitation, assaultive behavior, or criminal intent, and discusses appropriate responses

### *Written program*

A written program for job safety and security, incorporated into the organization's overall safety and health program, offers an effective approach for larger organizations. In smaller establishments, the program need not be written or heavily documented to be satisfactory. What is needed are clear goals and objectives to prevent workplace violence suitable for the size and complexity of the workplace operation and adaptable to specific situations in each establishment. The prevention program and startup date must be communicated to all employees. At a minimum, workplace violence prevention programs should do the following:

- Create and disseminate a clear policy of zero-tolerance for workplace violence, verbal and nonverbal threats, and related actions. Managers, supervisors, co-workers, and customers must be advised of this policy.

- Ensure that no reprisals are taken against an employee who reports or experiences workplace violence.
- Encourage employees to promptly report incidents and to suggest ways to reduce or eliminate risks; require records of incidents to assess risk and to measure progress.
- Outline a comprehensive plan for maintaining security in the workplace which includes establishing a liaison with law enforcement representatives and others who can help identify ways to prevent and mitigate workplace violence.
- Assign responsibility and authority for the program to individuals or teams with appropriate training and skills. The written plan should ensure that there are adequate resources available for this effort and that the team or responsible individuals develop expertise on workplace violence prevention in night retail establishments.
- Affirm management commitment to a worker-supportive environment that places as much importance on employee safety and health as on serving the customer.
- Set up a company briefing as part of the initial effort to address such issues as preserving safety, supporting affected employees, and facilitating recovery.

## *Worksite analysis*

Worksite analysis involves a step-by-step, common-sense look at the workplace to find existing or potential hazards for workplace violence. This entails reviewing specific procedures or operations that contribute to hazards and specific locales where hazards may develop.

A "Threat Assessment Team," similar task force, or coordinator may assess the vulnerability to workplace violence and determine the appropriate preventive actions to be taken. Implementing the workplace violence prevention program then may be assigned to this group. The team should include representatives from senior management, operations, employee assistance, security, occupational safety and health, legal, and human resources staff. The team or coordinator can review injury and illness records and workers' compensation claims to identify patterns of assaults that could be prevented by workplace adaptation, procedural changes, or employee training. As the team or coordinator identifies appropriate controls, these should be instituted. The recommended program for worksite analysis includes, but is not limited to, analyzing and tracking records, monitoring trends and analyzing incidents, screening surveys, and analyzing workplace security.

### *Records analysis and tracking*

This activity should include reviewing medical, safety, workers' compensation, and insurance records — including the OSHA 200 log, if required —

to pinpoint instances of workplace violence. Scan unit logs and employee and police reports of incidents or near-incidents of assaultive behavior to identify and analyze trends in assaults relative to particular departments, units, job titles, unit activities, work stations, and/or time of day. Tabulate these data to target the frequency and severity of incidents to establish a baseline for measuring improvement.

### *Monitoring trends and analyzing incidents*

Contacting similar local businesses, trade associations, and community and civic groups is one way to learn about their experiences with workplace violence and to help identify trends. Use several years of data, if possible, to trace trends of injuries and incidents of actual or potential workplace violence. Identify and analyze any apparent trends in injuries or incidents relating to a particular worksite, job title, activity, or time of day or week. The following questions may be helpful in ascertaining trends:

- How many times have your establishment or similar company establishments been robbed in the past three years? Have other violent incidents occurred? Analyze incidents, including noting characteristics of assailants and victims, a brief account of what happened before and during the incident, relevant details of the situation, and its outcome.
- How many injuries occurred during those robberies or other incidents?
- How many times was a firearm used?
- How many times was a firearm discharged?
- How many times was the threat of a firearm used?
- How many times were other weapons used?
- What job or workstation was involved in the robbery/incident? Identify, based upon the risk factors identified in these guidelines, those work positions in which staff are at risk of assaultive behavior.
- What specific tasks were employees performing prior to the robbery/incident? Identify processes and procedures that put employees at risk of assault. When do these occur (e.g., on all shifts)?
- What time was the robbery/incident committed?
- How many other incidents of violent attack or threats of violence on employees have been reported?
- How many times have police been called to your establishment to arrest a perpetrator, deter what you considered criminal intentions, or protect property? When possible, obtain reports of police who investigated the incident and their recommendations.

### *Screening surveys*

One important screening tool is giving employees a questionnaire or survey to get their ideas on the potential for violent incidents and to identify or confirm the need for improved security measures. Detailed baseline screening surveys can help pinpoint tasks that put employees at risk. Periodic surveys

— conducted at least annually or whenever operations change or incidents of workplace violence occur — help identify new or previously unnoticed risk factors and deficiencies or failures in work practices, procedures, or controls. Also, surveys help assess the effects of changes in the work processes. The periodic review process should also include feedback and follow-up. Independent reviewers, such as safety and health professionals, law enforcement or security specialists, insurance safety auditors, and other qualified persons may offer advice to strengthen programs. These experts also can provide fresh perspectives to improve a violence prevention program.

## *Workplace security analysis*

The team or coordinator should periodically inspect the workplace and evaluate employee tasks to identify hazards, conditions, operations, and situations that could lead to violence. To find areas requiring further evaluation, the team or coordinator should do the following:

- Analyze incidents, including the characteristics of assailants and victims, and give an account of what happened before and during the incident, as well as the relevant details of the situation and its outcome. When possible, obtain police reports and recommendations.
- Identify jobs or locations with the greatest risk of violence as well as processes and procedures that put employees at risk of assault, including how often and when.
- Note high-risk factors such as: physical risk factors of the building; isolated locations/job activities; lighting problems; lack of phones and other communication devices; areas of easy, unsecured access; and areas with previous security problems.
- Evaluate the effectiveness of existing security measures, including engineering, administrative, or other control measures. Determine if risk factors have been reduced or eliminated, and take appropriate action.

The following questions can help assess risks:

- Do employees exchange money with the public or guard valuable property or possessions during evening or late-night hours of operation?
- Is cash control a key element of the establishment's violence and robbery prevention program?
- Is there a drop safe to minimize cash on hand?
- Does the site have a policy to maintain less than $50 in the cash register?
- Are there signs posted stating that limited cash is on hand?
- Does the site have a policy limiting the number of cash registers open during late-night hours?

- Do employees receive training in emergency procedures for robberies, in conflict resolution, and in nonviolent response?
- At sites with a history of robbery or assaults, do employees have bullet-proof barriers or enclosures? Is more than one employee on duty?

After conducting a worksite analysis appropriate for the size and conditions of the workplace, the employer may find that there are no significant hazards related to violence in the establishment. If there are no hazards, the employer need not implement the other program elements recommended by the guidelines. The employer should, however, continue efforts to ensure workplace safety and health, monitoring changes in the workplace that might indicate hazards related to violence.

## *Hazard prevention and control*

After hazards of violence are identified through the systematic worksite analysis, the next step is to design measures through engineering or administrative and work practices to prevent or control these hazards. If violence does occur, post-incidence response can be an important tool in preventing future incidents. The workplace violence prevention and control program should indicate specific engineering, administrative, and work practice controls and other appropriate interventions that address specific hazards at the worksite. Presented in the sections that follow are general recommendations for establishments that provide retail services and are open late at night. These illustrative examples cover engineering and workplace adaptation and administrative and work practice controls as recommendations to help employers prevent workplace violence.

### *Engineering controls and workplace adaptation*

Engineering controls remove the hazard from the workplace or create a barrier between the worker and the hazard. The primary hazard faced by employees working in a late-night retail establishment is assault during an armed robbery. It is essential, therefore, to find ways to protect workers from assaults involving guns or other weapons. Physical changes in the workplace that help eliminate or reduce these hazards might include using some or all of the following measures. The selection of any measure should be based upon the hazards identified in the workplace security analysis of each facility:

- Physical barriers such as bullet-proof enclosures between customers and employees provide the greatest protection for workers. Installing pass-through windows for customer transactions and limiting entry to authorized persons during certain hours of operation also limit risk. Doors used for deliveries should be locked when not in use.
- Mechanisms that permit employees to have a complete view of their surroundings such as convex mirrors, elevated vantage point, and

placement of customer service and cash register areas so they are clearly visible outside of the retail establishment serve as deterrents.
- Video surveillance equipment and closed-circuit television can increase the possibility of detection and apprehension of the criminal, thus deterring crime.
- Adequate outside lighting of the parking area and approach to the retail establishment during nighttime hours of operation enhances employee protection. Surveillance lighting to detect and observe pedestrian and vehicular entrances of the retail establishment also helps. Adequate lighting within and outside the establishment (refer to ANSI/IES RP7-1993) makes the store less appealing to a potential robber by making detection more likely.
- Speed bumps placed in traffic lanes used to exit drive-up windows can deter would-be criminals by reducing the chance for a quick escape.
- An unobstructed view to the street from the store, clear of shrubbery, trees, or any form of clutter that a criminal could use to hide can help protect employees.
- Cash-handling controls, including the use of locked drop safes and the posting of signs stating that limited cash is on hand, can also deter would-be robbers.
- Height markers on exit doors should be installed to help provide more complete descriptions of assailants.
- Strategically placed fences can control access to the store.
- Garbage areas and external walk-in freezers or refrigerators should be located so as to ensure the safety of employees who use them. There should be good visibility with no potential hiding places for assailants near these areas.

## *Administrative and work practice controls*

Administrative and work practice controls affect the way jobs or tasks are performed. An effective program of hazard prevention and control includes safe and proper work procedures that are understood and followed by managers, supervisors, and workers. Key elements for preventing workplace violence include proper work practices, regular monitoring and feedback, modifications, and enforcement of the program. The following examples illustrate work practices and administrative procedures that can help prevent workplace violence incidents:

### *Proper work practices*

A program establishing proper work practices should include appropriate training and practice time for employees. Following are some suggested work practices:

- Employees should wear conservative clothing (such as a company uniform) and be discouraged from wearing jewelry.

- Employees should not carry cash while on duty unless it is absolutely necessary.
- During evening and late-night hours of operation, cash levels should be kept to a minimal amount per cash register ($50 or less) to conduct business. Transactions with large bills (over $20) should be prohibited.
- Stores should adopt proper emergency procedures for employees to use in case of a robbery or security breach. These should include incident report forms which prompt for perpetrator information (e.g., sex, height, build, age, race) to be completed immediately following a violent event. Emergency telephone numbers should be accessible and the notification policy clearly established. All violent incidents should be reported to local police.
- Alarm systems, video surveillance equipment, drop safes or comparable devices, surveillance lighting, or other security devices in the establishment must be used and maintained properly.
- Any physical barriers and/or pass-through windows must be used correctly to provide effective deterrence to robbers.
- At sites with a history of robbery or assaults, employees should not be required to work alone.

*Monitoring and feedback*

Regular monitoring helps ensure that employees continue to use proper work practices. Monitoring should include review of the specific procedures in use and their effectiveness, including a determination of whether the procedures actually used are those specified in the hazard prevention and control program. The review should address any deficiencies, changes that have occurred, employee or customer complaints about lack of security, instances of violence or threats of violence, and whether corrective action is necessary. Giving regular, constructive feedback to employees helps to ensure their commitment to the prevention program.

*Adjustments and modification*

Monitoring may show a need to modify administrative and work practice controls. Such adjustments could include:

- Limiting or restricting customer access
- Increasing staffing levels
- Reducing the number of cashier positions
- Increasing security or surveillance
- Reducing the hours of operation

*Enforcement*

For an effective program, the employer should establish employee sanctions for those employees who chronically and/or purposefully violate administrative controls or work practices. An employee who has been properly trained and consulted after such a violation, but who continues to violate established written work practice, should be disciplined accordingly.

### *Post-incident response*

Post-incident response and evaluation are essential to an effective violence prevention program. All workplace violence programs should provide comprehensive treatment for victimized employees and employees who may be traumatized by witnessing a workplace violence incident. Injured staff should receive prompt treatment and psychological evaluation whenever an assault takes place, regardless of severity. Transportation of the injured to medical care should be provided if care is not available on-site. Victims of workplace violence suffer a variety of consequences in addition to their actual physical injuries. These include short- and long-term psychological trauma; fear of returning to work; changes in relationships with co-workers and family; feelings of incompetence, guilt, and powerlessness; and fear of criticism by supervisors or managers. Consequently, a strong follow-up program for these employees will help them not only to deal with these problems but also to prepare them to confront or prevent future incidents of violence. There are several types of assistance that can be incorporated into the post-incident response. For example, trauma-crisis counseling, critical incident stress debriefing, or employee assistance programs may be provided to assist victims. Certified employee assistance professionals, psychologists, psychiatrists, clinical nurse specialists, or social workers could provide this counseling, or the employer can refer staff victims to an outside specialist. In addition, an employee counseling service, peer counseling, or support groups may be established. In any case, counselors must be well trained and have a good understanding of the issues and consequences of assaults and other aggressive, violent behavior. Appropriate and promptly rendered post-incident debriefings and counseling reduce acute psychological trauma and general stress levels among victims and witnesses. In addition, such counseling educates staff about workplace violence and positively influences workplace and organizational cultural norms to reduce trauma associated with future incidents.

## *Training and education*

Training and education ensure that all staff are aware of potential security hazards and how to protect themselves and their co-workers through established policies and procedures. Training should cover robbery prevention, reactions, post-robbery strategies, conflict resolution, and personal safety.

### *General training*

Training programs should include all employees, especially those engaged in exchanging money with customers (e.g., sales clerks, waitresses, bartenders). Employees who may face safety and security hazards should receive formal instruction on the specific hazards associated with the unit or job and facility, including information on potential injuries, problems identified in the facility, and the methods to control the specific hazards. Training for employees should be repeated for each employee as necessary, at least annually. In large establishments, refresher programs may be needed more frequently (monthly or quarterly) to effectively reach and inform all employees. New and returning employees should receive appropriate training through the establishment's violence prevention program. Employees should thoroughly understand the preventive measures and safeguards used at the specific worksite. Employees reassigned to jobs or tasks that increase the risk of potential violence (e.g., serving as cashier, making bank deposits, dealing with irate customers) should have specific training that covers these situations.

Training should be designed and implemented by qualified persons. Appropriate special training should be provided for personnel responsible for administering the program, and training should be presented in language and comprehension levels appropriate for the individuals being trained. It should provide an overview of the potential risk of assault, the prevention measures used to deter robbery or other assaults, the behavioral skills necessary to reduce the likelihood of a violent outcome, and the appropriate steps to take in case of an emergency. The training program also should include an evaluation component. This might include supervisor and/or employee interviews, testing and observing work practices in use, and reviewing actual incident reports of assaultive behavior.

### *Training for supervisors, managers, and security personnel*

Supervisors and managers are responsible for ensuring that employees follow safe work practices and receive appropriate training to accomplish this goal. Therefore, management personnel should undergo training comparable to that of the employees, plus additional training to enable them to recognize, analyze, and establish violence prevention controls. Training for managers should address any specific duties and responsibilities they have that could increase their risk of assault. Security personnel need specific training, including the psychological components of handling aggressive and abusive customers and ways to handle aggression and defuse hostile situations. The training program should also include an evaluation. The content, methods, and frequency of training should be reviewed and evaluated annually by the team or coordinator responsible for implementation. Program evaluation may involve supervisor and/or employee interviews, testing, and observing and/or reviewing reports of behavior of individuals in threatening situations.

## *Recordkeeping and evaluation*

Recordkeeping and evaluation of the violence prevention program are necessary to determine overall effectiveness and identify any deficiencies or changes that should be made.

## *Recordkeeping*

Recordkeeping is essential to the success of a workplace violence prevention program. Good records help employers determine the severity of the problem, evaluate methods of hazard control, and identify training needs. Records can be especially useful to large organizations and for members of a business group or trade association who "pool" data. Records of injuries, illnesses, accidents, assaults, hazards, corrective actions, and training, among others, can help identify problems and solutions for an effective program. The following records are important:

- OSHA regulations require entry on the Injury and Illness Log (OSHA 200) of any injury that requires more than first-aid, is a lost-time injury, requires modified duty, or causes loss of consciousness. (This applies only to establishments required to keep OSHA logs). Injuries caused by assaults, which are otherwise recordable, also must be entered on the log. A fatality or catastrophe that results in the hospitalization of three or more employees must be reported to OSHA within 8 hours. This includes those resulting from workplace violence and applies to all establishments.
- Medical reports of worker injury and supervisors' reports for each recorded assault should describe the type of assault (i.e., unprovoked sudden attack), who was assaulted, and all other circumstances of the incident. The records should include a description of the environment or location, potential or actual cost, lost time, and the nature of injuries sustained.
- Incidents of abuse, verbal attacks or aggressive behavior — which may be threatening to the worker but do not result in injury, such as pushing or shouting and acts of aggression towards customers — should be recorded, perhaps as part of an assaultive incident report. These reports should be evaluated routinely by the affected department.
- Minutes of safety meetings, records of hazard analyses, and corrective actions recommended and taken should be documented.
- Records of all training programs, attendees, and qualifications of trainers should be maintained.

## *Evaluation*

As part of their overall program, employers should evaluate their safety and security measures. Top management should review the program regularly, and with each incident, to evaluate program success. Responsible parties (managers, supervisors, and employees) should collectively re-evaluate policies and procedures on a regular basis. Deficiencies should be identified and corrective action taken. An evaluation program should involve the following:

- Establishing a uniform violence reporting system and regular review of reports
- Reviewing reports and minutes from staff meetings on safety and security issues
- Analyzing trends and rates in illness/injury or fatalities caused by violence relative to initial or "baseline" rates
- Measuring improvement based on lowering the frequency and severity of workplace violence
- Keeping up-to-date records of administrative and work practice changes to prevent workplace violence to evaluate their effectiveness
- Surveying employees before and after making job or worksite changes or installing security measures or new systems to determine their effectiveness
- Keeping abreast of new strategies available to deal with violence in the night retail industry as these develop
- Surveying employees who experience hostile situations about the medical treatment they received initially and, again, several weeks afterward, and then several months later
- Complying with OSHA and state requirements for recording and reporting deaths, injuries, and illnesses
- Requesting periodic law enforcement or outside consultant review of the worksite for recommendations on improving employee safety

Management should share workplace violence prevention program evaluation reports with all employees. Any changes in the program should be discussed at regular meetings of the safety committee, union representatives, or other employee groups.

## *Sources of assistance*

Employers who would like assistance in implementing an appropriate workplace violence prevention program can turn to the OSHA consultation service provided in their state. Primarily targeted at smaller companies, the consultation service is provided at no charge to the employer and is independent of OSHA's enforcement activity.

OSHA's efforts to assist employers combat workplace violence are complemented by those of NIOSH (1-800-35-NIOSH) and public safety officials, trade associations, unions, insurers, and human resource and employee assistance professionals, as well as other interested groups. Employers and employees may contact these groups for additional advice and information.

## *Conclusions*

The Occupational Safety and Health Administration recognizes the importance of effective safety and health program management in providing safe and healthful workplaces. In fact, OSHA's consultation services help employers establish and maintain safe and healthful workplaces, and the agency's Voluntary Protection Program was specifically established to recognize worksites with exemplary safety and health programs. Effective programs are known to improve both morale and productivity and reduce workers' compensation costs. OSHA's violence prevention guidelines are an essential component to workplace safety and health programs. OSHA believes that the performance-oriented approach of the guidelines provide employers with flexibility in their efforts to maintain safe and healthful working conditions.

*appendix H*

# *Machine guarding checklist*

Answering the following questions should help the interested reader determine the safeguarding needs of his/her own workplace by drawing attention to hazardous conditions or practices requiring correction.

## *Requirements for all safeguards*

| | | Yes | No |
|---|---|---|---|
| 1. | Do the safeguards provided meet the minimum OSHA requirements? | ____ | ____ |
| 2. | Do the safeguards prevent workers' hands, arms, and other body parts from making contact with dangerous moving parts? | ____ | ____ |
| 3. | Are the safeguards firmly secured and not easily removable? | ____ | ____ |
| 4. | Do the safeguards ensure that no object will fall into the moving parts? | ____ | ____ |
| 5. | Do the safeguards permit safe, comfortable, and relatively easy operation of the machine? | ____ | ____ |
| 6. | Can the machine be oiled without removing the safeguard? | ____ | ____ |
| 7. | Is there a system for shutting down the machinery before safeguards are removed? | ____ | ____ |
| 8. | Can the existing safeguards be improved? | ____ | ____ |

## Mechanical hazards

### Point of operation

| | | Yes | No |
|---|---|---|---|
| 1. | Is there a point-of-operation safeguard provided for the machine? | ____ | ____ |
| 2. | Does it keep the operator's hands, fingers, and body out of the danger area? | ____ | ____ |
| 3. | Is there evidence that the safeguards have been tampered with or removed? | ____ | ____ |
| 4. | Could you suggest a more practical, effective safeguard? | ____ | ____ |
| 5. | Could changes be made on the machine to eliminate the point-of-operation hazard entirely? | ____ | ____ |

### Power transmission apparatus

| | | Yes | No |
|---|---|---|---|
| 1. | Are there any unguarded gears, sprockets, pulleys, or flywheels on the apparatus? | ____ | ____ |
| 2. | Are there any exposed belts or chain drivers? | ____ | ____ |
| 3. | Are there any exposed set screws, key ways, collars, etc.? | ____ | ____ |
| 4. | Are starting and stopping controls within easy reach of the operator? | ____ | ____ |
| 5. | If there is more than one operator, are separate controls provided? | ____ | ____ |

### Other moving parts

| | | Yes | No |
|---|---|---|---|
| 1. | Are safeguards provided for all hazardous moving parts of the machine, including auxiliary parts? | ____ | ____ |

## Nonmechanical hazards

| | | Yes | No |
|---|---|---|---|
| 1. | Have appropriate measures been taken to safeguard workers against noise hazards? | ____ | ____ |
| 2. | Have special guards, enclosures, or personal protective equipment been provided, where necessary, to protect workers from exposure to harmful substances used in machine operation? | ____ | ____ |

## *Electrical hazards*

| | | Yes | No |
|---|---|---|---|
| 1. | Is the machine installed in accordance with National Fire Protection Association and National Electrical Code requirements? | ____ | ____ |
| 2. | Are there loose conduit fittings? | ____ | ____ |
| 3. | Is the machine properly grounded? | ____ | ____ |
| 4. | Is the power supply correctly fused and protected? | ____ | ____ |
| 5. | Do workers occasionally receive minor shocks while operating any of the machine? | ____ | ____ |

## *Training*

| | | Yes | No |
|---|---|---|---|
| 1. | Do operators and maintenance workers have the necessary training in how to use the safeguards and why? | ____ | ____ |
| 2. | Have operators and maintenance workers been trained in where the safeguards are located, how they provide protection, and what hazards they protect against? | ____ | ____ |
| 3. | Have operators and maintenance workers been trained in how and under what circumstances guards can be removed? | ____ | ____ |
| 4. | Have workers been trained in the procedures to follow if they notice guards that are damaged, missing, or inadequate? | ____ | ____ |

## *Protective equipment and proper clothing*

| | | Yes | No |
|---|---|---|---|
| 1. | Is protective equipment required? | ____ | ____ |
| 2. | If protective equipment is required, is it appropriate for the job, in good condition, kept clean and sanitary, and stored carefully when not in use? | | |
| 3. | Is the operator dressed safely for the job (e.g., no loose-fitting clothing or jewelry)? | ____ | ____ |

## *Machinery maintenance and repair*

| | | Yes | No |
|---|---|---|---|
| 1. | Have maintenance workers received up-to-date instruction on the machines they service? | ___ | ___ |
| 2. | Do maintenance workers lock out the machine from its power sources before beginning repairs? | ___ | ___ |
| 3. | Where several maintenance persons work on the same machine, are multiple lockout devices used? | ___ | ___ |
| 4. | Do maintenance personnel use appropriate and safe equipment during their repair work? | | |
| 5. | Is the maintenance equipment itself properly guarded? | | |
| 6. | Are maintenance and servicing workers trained in the requirements of 29 C.F.R. 1910.147 (lockout/tagout of hazards), and do the procedures for lockout/tagout exist before they attempt their tasks? | ___ | ___ |

# *Index*

# Index


## A

Administrative controls, 29, 43, 51, 60
Administrative Law Judge (ALJ), 17, 18
Administrative Procedure Act, 5
Air-purifying respirator (APR), 63
Airborne, 62
American Lung Association, 63
American National Standards Institute (ANSI), 5, 6
Anchorage, 108
Appeal, 18
Asbestosis, 62
Asphyxiants, 61
Assembly line, 7
Atmospheric testing/monitoring, 84
Audiograms, 120–122
Audiometers, 123
Audiometric testing, 120, 125
- Purpose, 124
- Requirements, 122–123

Auditing, 31
Automatic suppression systems, 76–77

## B

Bending, 104
Bhopal, 187
Biohazard, 129
Black lung, 62
Bloodborne pathogen standard, 127–129, 168–170
- Definitions, 131–133
- Manual, 134
- Program, 130, 172–174

Body belt/harness, 108
Buckle, 108
Buzz word, 25

## C

Calibrate, 61
Carbon monoxide, 61
Carcinogens, 59
Carpal tunnel syndrome, 26
Chemical cartridge respirator, 64
Chemical hazards, 183–185
- History of, 183

Chronic obstructed pulmonary disease (COPD), 62
Citations, 17
Code of Federal Regulations, 33
- Part 1910, 33–35

Communication problems, 83–84
Complainant, 18
Compliance,
- Deadlines, 50
- Lockout/tagout, 99
- Officer, 7, 16

Comply, 5
Conditioning period, 28
Confined space, 60
- Defined, 80

Entry rescue, 86
Hazards, 79–80
Permit required, 85–86
Rescue priorities, 86–87
Training, 87–88
Written program elements, 88–89
Congress, 14
Connector, 108
Constitutional rights, 19
Contaminated sharps, 131, 146–147
Controlled access zone (CAZ), 108
Corrosive hazards, 81
Cover-up, 15
Cumulative trauma disorder (CTD), 26
Cutting, 103

## D

Dangerous equipment, 108
Deceleration device, 108
Decibels, 124
Decontamination, 131, 148
Definitions, 51–58
De minimis, 8
Department of Labor, 13–15
Design,
Tool/handle, 28
Work methods, 28
Direct review, 18
Disciplinary action, 32
Discrimination, 17

## E

Elasticity, 62
Electrical safety, 91–95
Effects of current, 92
Enforcement, 95
General operating rules, 92–93
Planning, 93
Safety checklist, 95
Safety equipment, 93–94
Emergency, disaster preparedness, 175–176
Media relations, 176–179
Planning, 177
Risk assessment, 176
Shareholder, 179–181
Emergency, temporary standard, 6
Emergency Planning/Community Right-To-Know Act (EPCRA), 187, 188–190
Energy hazards, 81
Energy-isolating device, 99–100
Enforcement, 15
Engineering controls, 27, 43, 52, 60, 132, 136–140
Engulfment (contents) hazards, 82
Environmental, 16, 63, 146
Environmental condition hazards, 83
Environmental Protection Agency (EPA), 62
Equivalent, 108
Ergonomic, 25
Checklist, 32–33
Define, 25–26
Mandatory guidelines, 25
Proposed standard, 33–35
Risk factors, 40–43, 52
Six elements, 36
Team, 26
Workplace, 25
Exposure Control Plan, 133, 184
Exposure incident, 132

## F

Fail-safe, 105
Failure to abate, 8, 12–13
Fall arrest, 110, 111
Equipment, 111
Fall hazards, 82, 107
Countermeasures, 107
Factors, 107
Fall protection, 108
Categories, 111
Duty, 112–113
Equipment definitions, 108–111, 112
Systems, 112, 110, 112
Training, 113–114
Fan system, 60
Federal Emergency Management Agency, 71
Federal Register, 5–6, 183
Fire detection methods,, 76
Fire/explosion hazard, 71
National fire problem, 72
Overall picture, 71

Fire/explosive atmospheric hazards, 82
Fire prevention, 76
Fire strategy, 74
  Six steps, 74–78
Fit testing, 69
Follow-up, 53
Free fall, 109
  Distance, 109

## G

General duty clause, 6, 8, 25
Gravity, 7
Guardrail system, 109

## H

Have knowledge, 53
Hazard(s),
  Chemical, 183
  Classification, 80–82
  Control, 27, 43–44, 60
  Fall, 107
  Information, 39
  Prevention, 27
  Reporting, 39–40
Hazard Communication Standard, 185–186
Hazardous energy, 96
  Control, 96
  Lockout/tagout standard, 96–97
HBV, 132, 153
Healthcare professional (HCP), 46, 53
Hearing conservation, 117
  Appendices, 126
  Exemptions, 126
  Monitoring, 119
  Programs, 118–119, 120, 124
  Protectors, 123
  Recordkeeping, 125
  Requirements, 122–123
  Sound levels, 117–118
Hearing protection, 115–116
  Training topics, 116
  Written program, 116
Hepatitis, 150–157, 160, 162, 165
HIV, 132, 153, 161
Hole, 109
Human resources (HR) professionals, 12

## I

Ice bullets, 63
Incremental abatement process, 44
Inert gases, 61
Infeasible, 109
Informal rulemaking, 5
Inspection(s)
  Document forms, 16
  OSHA checklist, 18–19
  OSHA, four components, 16
  Warrant, 15
Insurance Service Office (ISO), 75
Interior structural brigades, 78
International Fire Service Training Association, 66
International Society of Fire Service Instructors, 66
Interstate commerce, 3
Investigation, reasonable investigative techniques, 16

## K

Knowing the rules, 20

## L

Lanyard, 109
  Self-retracting, 110
Leading edge, 109
Liabilities, 8
  Criminal, 13–15
Lifeline, 109
  Self-retracting, 110
Lockout/tagout standard, 96
  Four stages, 96–97
  Procedures, 97–98
  Production operations, 98–99
  Sequence of procedure, 98
  Training, 99–100
Logical outgrowth, 6
Low-slope roof, 109
Lower levels, 109

## M

Machine guarding, 101
  General requirements, 101

Maintenance,
Machine/equipment, 105
Machine guarding checklist, 106
Safeguards, 105–106
Management,
commitment, 26
leadership/employee participation, 38–39
team, 194
Mandate, 6
Manual handling jobs, 35, 53
Manufacturing jobs, 35, 54
Material Safety Data Sheet (MSDS), 51, 185
Eliminate, 44, 52
Hazards, 37–38, 55
Management, 46–48, 55–56
OSHA-recordable, 56
Quick Fix, 37–38
Signs/symptoms, 38–39, 56, 57
Mechanical equipment, 82, 109
Hazards, 82
Mechanical motions, 102
Mechanical power transmission, 101
Medical management, 29
Mesothelioma, 63
Methane, 61
Mine Safety and Health Administration (MSHA), 63
Miranda Warning, 19–20
Monetary, 7
Fine, 7
Liabilities, 8
Penalties, 8
Monitoring, 28, 119
Musculoskeletal disorder, 35, 55

## N

National Advisory Committee on Occupational Safety & Health (NACOSH), 6
National Consensus Standards, 5
National Electric Code (NEC), 5, 91
National Fire Data Center, 71
National fire problem, 72
Causes, 73
Nonresidential properties, 74
Six-steps strategy, 74–78
What saves lives, 73
Where fire occurs, 71–72
Who is at risk, 73
National Fire Protection Association (NFPA), 66, 75, 77
Code, 75
National Institute for Occupational Safety & Health (NIOSH), 4, 63, 183
Nonadversarial administrative hearing, 5
Notice, 15
Notice of contest, 17

## O

Occupational exposure, 132
Occupational illness, 10
Occupational Safety & Health Administration (OSHA), 4, 14, 15, 25, 62, 91, 108, 183
Office of Publications, 67
Occupational Safety & Health Review Commission (OSHRC), 4, 17, 18
Omnibus Budget Reconciliation Act, 6
Violation and Penalty Schedule, 7
On the clock, 56
Opening, 109
OSH Act, 3, 10, 13, 60, 77–78
OSHA Bloodborne Pathogen Standard, 127–129
OSHA Compliance Field Operations Manual (OSHA Manual), 9, 12–13
OSHA Hearing Conservation Standard, 116
OSHA-recordable MSD, 36, 56
Overhand bricklaying, 109

## P

Parenteral, 132
Particulates, 62
PDR (Petition for Discretionary Review), 18
Penalty table, 7
Periodically, 57
Permit required, 80
Personal protective equipment (PPE), 29, 43–44, 57, 59, 60, 94, 106, 132, 141–146
Physical work activities, 57

PMA (Petition Modification Abatement), 17
Point of operation, 101–102
Positioning device, 110, 111
Power transmission, 102
PRCS Entry Permit, 89
Problem job, 57
Process actions, 103
Program evaluation, 48–49
Public hearing, 5
Pull the ticket, 4
Pulsed tone, 123
Punching, 103–104

## Q

Qualitative fit testing (QLFT), 64
Quantitative fit testing (QNFT), 64, 65

## R

Reasonable promptness, 17
Reciprocating, 102–103
Records, 49
Recovery period, 47
Regulatory issues, 3
Repeat violations, 8, 12,13
Resources, 57
Respirators, 63
  Types, 63–65
  Users, 65–66
Respiratory hazards, 59, 80–81
  Classification, 61,
  Exposure prevention, 60
  Factor, 59
  Gases/fumes, 61
Respiratory Protection Program, 67
Respondent, 18
Retrieval plan, 112
Revoke, 5
Right-To-Know Standard, 183–184, 187
Roof, 110
Rope grab, 110
Rotating, 102–103
Rulemaking, 5

## S

Safety
  Device failure, 104–105
  Monitoring system, 110
  Safeguards, 105–106
Secretary of Labor, 4–6, 17
Self-contained breathing apparatus (SCBA), 61, 64
  Closed-/open-circuit, 64–65
Serious physical harm, 10
Shearing, 104
SIC Code, 186–187
Silica, 62
Small Entity Compliance Guide, 67
Snaphook, 110
Sorbent cartridges, 63
Sound level contours, 117–118
Spot analyzers, 62
Standard threshold shift, 122
Standards, 5
  Emergency temporary, 6
  Permanent, 6
State plan, 4, 15
Statute of Limitations, 14, 17
Stay, 17
Steep roof, 110
Stock, 180
Superfund Amendment Re-Authorization Act (SARA), 187
Supplied-air respirators (SAR), 64
Suspension system, 111–112
Symptoms survey
  Checklist, 30
  Define, 30

## T

Tendinitis, 26
Tennis elbow, 26
Toxic vapors/gases, 60, 61
Track, 180
Training/education programs, 30, 44–46
  Certification, 114
  Fall protection, 113–114
  Five P's, 66
  General vs. job-specific, 31
  Maintenance, 31
  Respirator users, 65–66
  Retraining, 114
  Supervisors/managers, 31
Transversing, 102–103
Trauma disorders, 58
Trigger finger, 26

## U

U.S. Constitution, 15
U.S. Court of Appeals, 18
U.S. Department of Justice, 8, 14
U.S. Fire Administration, 71

## V

Violation(s), 7, 9
  Criminal, 14
  De minimis, 8–9
  Four-step approach, 11
  Infraction, 7
  Nonserious, 8–10
  Notices, 13
  Repeat, 8, 12–13
  Serious, 7, 8–10
  Willful, 8, 11–12

## W

Walking/working surface, 110
Wall Street, 180
Warning device, 99
  Line system, 110
Website analysis, 26
Willful violations, 8
Work
  Area, 110
  Practice controls, 28, 43, 57
  Restrictions, 46–47, 58
  Techniques, 28
Work Restriction Protection (WRP), 46, 57–58
Workplace
  Deaths, 15
  Periodic walkthrough, 30, 47
Workplace violence, 191–198
  Crisis plan, 196–197
  Investigation format, 195
  Management team, 194
  Prevention program, 193, 196
  Security, 196
  Threats, 197
  Warning signs, 191–192, 193
Worksite analysis, 27
Workstation design, 27
Written confined space program, 88–89
Written opinion, 46–47
Written respirator program, 67
  Elements, 67–69
  Evaluation factors, 68–69
  Fit testing, 69
  Medical evaluation, 69
  Recordkeeping, 69